T0313787

69th Porcelain Enamel Institute Technical Forum

69th Porcelain Enamel Institute Technical Forum

A Collection of Papers Presented at the 69th
Porcelain Enamel Institute Technical Forum
September 17–20, 2007
Indianapolis, Indiana

Conference Director
Holger Evele

Assistant Conference Director
Peter Vodak

Editor
William D. Faust

A John Wiley & Sons, Inc., Publication

Published by John Wiley & Sons, Inc., Hoboken, New Jersey.
Published simultaneously in Canada.

For general information on our other products and services or for technical support, please contact our
Customer Care Department within the United States at (800) 762-2974, outside the United States at
(317) 572-3993 or fax (317) 572-4002.

Wiley also publishes its books in a variety of electronic formats. Some content that appears in print may
not be available in electronic format. For information about Wiley products, visit our web site at
www.wiley.com.

Library of Congress Cataloging-in-Publication Data is available.

ISBN 978-0-470-19641-0

Printed in the United States of America.

10 9 8 7 6 5 4 3 2 1

Contents

Preface

The Porcelain Enamel Institute Technical Forum Committee is pleased to present the proceedings of the 69th Annual PEI Technical Forum. This volume culminates a year's worth of work by the committee and our presenters. The three day's of meetings, seminars and presentations was held in the Indianapolis Convention Center September 17-20, 2007 in conjunction with the FIN-X'07 International Expo and Conference for Industrial Finishers. Work has already started on the 70th Technical forum to be held in Nashville, TN June 9-12, 2008 at the downtown Doubletree Hotel.

I would like to thank the entire committee, especially my vice-chairman and incoming chairman Peter Vodak (Engineered Storage Products) for their time, efforts and support in making this forum as successful as the previous 68. Without the participation of everyone the forum could not be the success it was. Each year we strive to bring to the industry reports of new technology available and other technical presentations to support our industry, mission accomplished.

As the Technical Forum brings new timely information to the industry our Back to Basics seminar provides a training opportunity for both industry newcomers and individuals seeking refresher training the chance to learn from industry experts. This year was again well attended and well received. Thanks to all the faculty for their efforts in preparing and presenting fundamental information to wide range of eager students.

I and the committee would like to thank the authors and presenters of the may topics covered during the technical forum. We and the industry are thankful for the time and effort in researching, preparing and presenting their informative papers. We also want to thank those who took time off their busy schedules to attend the Technical Forum. Their attention and discussion increased the value for everyone at the forum.

Our final thanks goes to William "Darry" Faust, our compiler and editor of these proceeding for many years. Darry has chosen to retire. It will difficult to find someone as resourceful and dedicated to put future proceedings together.

Peter Vodak, as the incoming Chairman, along with Mike Horton, new Vice-

Chairman, and the entire committee has already begun preparations for the next Technical Forum.

It promises to be another informative and worthwhile meeting.

HOLGER EVELE
Ferro Corporation
Chariman 2007 PEI Technical Forum Committee and Back-to-Basics Seminar

2007 PEI Officers

Chairman of the Board
JACK MCMAHON
Pemco Corporation

President
BOB HARRIS
Hanson Industries

Vice Presidents
BILL GANZER
Mapes & Sprowl Steel

KEN KREEGER
Nordson Company

DON MCCORMICK
Electrolux Home Products

TIM SCOTT
Henkel Surface Technologies

NICK SEDELIA
Whirlpool Corporation

MILES VOLTAVA
Ferro Corporation

PAT WALSH
Porcelain Industries

2007 Technical Forum Committee

Chairman: Holger Evele, Ferro Corporation
Vice Chairman: Peter Vodak, Engineered Storage Products

Peter Dority, Coral Chemical Company
Cullen Hackler, Porcelain Enamel Institute
Mike Horton, KMI Systems
Ken Kreeger, Nordson Corporation
Liam O'Byrne, O'Byrne Consulting Services
Tim Scott, Henkel Surface Technologies
Larry Steele, Mapes & Sprowl Steel
Dave Thomas, American Trim
Miles Votava, Ferro Corporation
Jack Waggener, URS Corporation
Mike Wilczynski, A O Smith Protective Coatings
Ted Wolowicz, Electrolux Home Products
Jeff Wright, Ferro Corporation
David Latimer, Whirlpool Corporation

Porcelain Enamel Institute
PO Box 920220
Norcross, GA 30010
Phone : 770-281-8980
E-mail : penamel@aol.com
www.buyporcelain.org
www.porcelainenamel.com

Past Chairs of PEI Technical Forums

Holger Evele	2006–07
Ferro Corporation	
Steve Kilczewski	2004–05
Pemco Corporation	
Liam O'Byrne	2002–03
AB&I Foundry	
Jeff Sellins	2000–01
Maytag Cooking Products	
Robert Reese	1998–99
Frigidaire Home Products	
David Thomas	1996–97
The Erie Ceramic Arts Company	
Rusty Rarey	1994–95
LTV Steel Company	
Douglas Giese	1992–93
GE Appliances	
Anthony Mazzuca	1990–91
Mobay Corporation	
William McClure	1988–89
Magic Chef	
Larry Steele	1986–87
Armco Steel	
Donald Sauder	1984–85
WCI–Range Division	
James Quigley	1982–83
Ferro Corporation	
George Hughes	1980–81
Vitreous Steel Products Company	
Lester Smith	1978–79
Porcelain Metals Corporation	

Evan Oliver   1977
Bethlehem Steel Corporation
Wayne Gasper   1975–76
The Maytag Company
Donald Toland   1973–74
U.S. Steel Corporation
Archie Farr   1971–72
O. Hommel Company
Harold Wilson   1969–70
Vitreous Steel Products Company
Forrest Nelson   1967–68
A.O. Smith Corporation
Grant Miller   1965–66
Ferro Corporation
Mel Gibbs   1963–64
Inland Steel Company
Charles Kleinhans   1961–62
Porcelain Metals Corporation
James Willis   1959–60
Pemco Corporation
Lewis Farrow   1957–58
Whirlpool Corporation
Gene Howe   1955–56
Chicago Vitreous Corporation
W.H. "Red" Pfieffer   1953–54
Frigidaire Division, G.M.C.
Roger Fellows   1951–52
Chicago Vitreous Corporation
Glenn McIntyre   1948–50
Ferro Corporation
Frank Hodek   1936–47
General Porcelain Enameling and Mfg. Company

SURFACE TENSION AND FUSION PROPERTIES OF PORCELAIN ENAMELS

Charles Baldwin and Sid Feldman
Ferro Corporation

ABSTRACT

High-temperature viscosity of glasses is an important key to understanding the phenomena that occur when firing porcelain enamels.

PORCELAIN ENAMELS

Porcelain enamel is a glass coating fused to metal. While there are similarities between the spray application of enamels and paints, there are important differences because of the vitreous nature of porcelain.

Porcelain enamel glass frit is typically alkali borosilicate where the alkalis are lithium, sodium, or potassium. Alkaline-earth ions such as calcium, strontium, or barium are also often present as fluxes. For adhesion, cobalt, nickel, iron, and copper are added to the glass.

During smelting, the molten glass is quenched, which results in an amorphous molecular structure. Glass does not have a distinct melting point, which is a sudden transition from a solid to liquid phase, but instead has a glass temperature (T_g) at which the solid glass becomes a highly viscous supercooled liquid. As the temperature is increased above T_g, the viscosity drops.

GLASS VISCOSITY

Viscosity is resistance of a fluid to deformation. For a glass, the viscosity varies with temperature according to the Vogel-Fulcher-Tamman (VFT) equation[1]:

$$\log (n) = A + B/(T-T_0)$$

This is equivalent to the Williams-Landel-Ferry (WLF) model for viscosity of a polymer above its glass temperature, but neither model is used in the enameling industry. Typically, the enamel industry has interpreted glass viscosity according to the type of generalized diagram in Figure 1. On the chart, the borosilicate and soda-lime glass pass through the various points on the viscosity scale at lower temperatures than the nearly pure silica glasses. The curve for porcelain enamel alkali borosilicate glasses would be expected to be lower and to the left of the one for borosilicate glass.

1

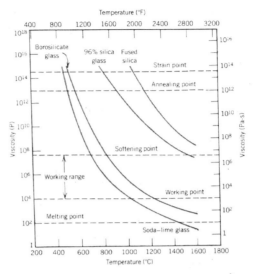

Figure 1. Glass viscosity versus temperature[2]

FIRING ENAMELS

It is most straightforward to consider the processes, which occur during firing of a ground coated part. First, the dried enamel and steel expand according to the respective thermal expansion (α) of each. As such, the value of α, T_g, and dilatometric softening point (T_s) of three ground coats was measured with an Orton Model 1000R dilatometer.

Samples bars for thermal expansion measurements were prepared by placing 8 g of glass frit powder into a carbon mold. The mold was fired at 1550°F (621°C) for 12 min. Next, the glass bar was placed into a 1000°F (538°C) furnace, which was switched off and allowed to cool overnight. Then, the bar was cut to a length of 2 ± 0.2 inches (50.8 ± 5 mm), and the edges were rounded on a grinding wheel.

The sample was placed in a quartz tube in contact with a ceramic pushrod. As the temperature was increased from room temperature, a linear voltage displacement transducer (LVDT) measured the expansion of the bar. When the temperature was reached that the bar viscosity was low enough to start contracting under the force of the pushrod, the dilatometer automatically shut off shortly after a preset amount of contraction. That temperature is the T_s.

The thermal expansion curves are shown in Figure 2. The theoretical linear curve for enameling steel based on $\alpha = 12.1 \times 10^{-6}/°C$ is included for reference. The value of α was determined by calculating the slope of the linear portion of the curve below 662°F (350°C) and is marked accordingly.

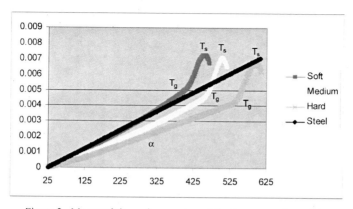

Figure 2. Measured thermal expansion of the three ground coat frits

While a number of sophisticated tests exist to measure glass viscosity versus temperature,[3] the fusion flow test is used by the enamel industry to determine relative glass viscosity at a fixed temperature. Two grams of -10/+30 M frit powder were pressed into a pellet, which was placed along a line on the top edge of a ground coated bent panel. The panel was put into a furnace at 1500°F (816°C) for 1.5 min and then tilted up. The fusion flow, f, is:

$$f = x_{sample} - x_{ref}$$

where x_{sample} is the distance from a line parallel to the top edge of the panel that the test sample flowed, and x_{ref} is the distance run by the reference standard. The fusion flow is negative if the test sample runs a shorter distance than the standard. The results are given in Table 1 with the values of α, T_g, and T_s.

Table 1. Thermal expansion and viscosity results

Glass	α (x 10^{-6}/°C)	T_g (°C)	T_s (°C)	Fusion Flow (816°C)
Soft	13.1	407	457	+170
Medium	10.0	456	504	-6
Hard	8.1	536	594	-25

ENAMEL WETTING ON OXIDIZED STEEL

As the metal temperature increases during firing, the reactions occur that create adhesion of the enamel to the steel. Additionally, good spreading of the liquid enamel on the steel is a basic requirement for adhesion. The balance of forces arising from a drop of liquid enamel (l) on surface (s) under a vapor (v) is schematically shown in Figure 3.

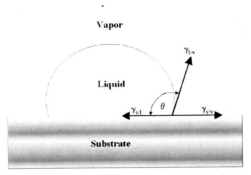

Figure 3. Liquid droplet on a solid surface[4]

Young's equation describes the balance of forces:

$$\gamma_{s/v} = \gamma_{s/l} + \gamma_{l/v} \cos\theta$$

where γ is the energy per unit area of the appropriate interface and θ is the contact angle between the liquid and the substrate. If $\gamma_{s/v} > \gamma_{s/l}$, the surface will be wetted to decrease the area of the higher energy s/v interface; this is the preferred situation. If $\gamma_{s/v} < \gamma_{s/l}$, balling up of the liquid will occur to reduce the area of the higher energy s/l interface.

With typical enamel firing temperature between 1400 to 1600°F (760 to 871°F), the reactivity of the steel affects the wetting, and high-temperature reactive wetting remains relatively poorly understood.[5] In fact, the liquid enamel is not in contact steel, but actually with iron oxide that forms after firing begins. K. Sarrazy showed this by firing an enamel (with $\alpha = 11.5 \times 10^{-6}/°C$, $T_g = 475°C$ (887°F), and $T_s = 545°C$ (1013°F)) on steel under argon and air as part of a study of adhesion. The appearance of the panels is shown in Figure 4. Since the iron oxide did not form under argon, $\gamma_{s/v} < \gamma_{s/l}$, the enamel could not wet the steel, and adhesion would not be possible.

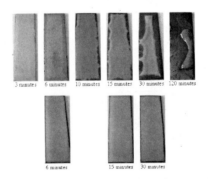

Figure 4. Clear enamel after firing at 1472°F (800°C) under argon (top) and air (bottom)[6]

The surface tension of the soft, medium, and hard frits was estimated by calculating a weighted average of the oxide components and is shown in

Table 2. It suggests that the soft frit will actually wet the substrate the least, and this was observed on the fusion test. The flow on the soft frit was narrower and at a greater contact angle than the medium frit.

Table 2. Estimated surface tension

Frit	Surface Tension (Dynes/cm)
Soft	2.74
Medium	2.46
Hard	2.32

Figure 5 shows the viscosity is a more important factor than the surface tension. The fusion flow buttons were melted on a steel plate and a high-density aluminum oxide tile at 1500°F (816°C). The tile was used as a non-reactive substrate. The soft frit wetted the steel and tile much more than the other two frits. Therefore, the glass viscosity at the firing temperature is a more important factor than the surface tension. On the steel panel, it should be noted that the hard frit shattered, presumably from tensile stresses that arose on cooling.

 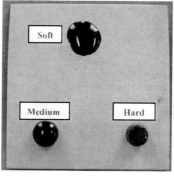

Figure 5. Button flow results on sheet steel (left) and alumina tile (right)

ENAMEL/STEEL REACTIONS

The bisque enamel permits the transport of O_2 from the surrounding air to the enamel/steel interface, resulting in the formation of an iron oxide (FeO) scale on the surface of the steel starting at about 600°F (316°C). According to Dietzel[7] and also King[8], cobalt or nickel precipitated from the ground coat glass in contact with the iron-containing substrate forms a short-circuited local cell in which iron is the anode. The current flows from the iron through the molten enamel to the cobalt and back to the iron. These local cells are not exhausted during firing because there is an abundance of anodic iron (in the steel), and diffusing atmospheric oxygen has a depolarizing action on the cathode side (the cobalt and nickel). The result is that the iron goes continuously into solution, the surface becomes roughened, and the glass anchors itself into the holes. The required galvanic reactions are:

1) $Fe^0 + CoO \rightarrow FeO \text{ (wustite)} + Co^0$
2) $2Co^0 + O_2 \rightarrow 2Co^{2+} + 2O^{2-}$
3) $Co^{2+} + 2e^- \rightarrow Co^0$
4) $Fe^0 + 2e^- \rightarrow Fe^{2-}$

On optical micrographs, the iron-rich layer is visible as a brown haze layer at the enamel/steel interface. Some evidence of the reaction reaching the enamel surface as copperheads is visible in the puddle of soft frit in

Figure 5. A cross-section of the reaction layer from Sarrazy's study is shown on a scanning electron microscope (SEM) micrograph in Figure 6.

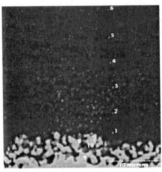

Figure 6. SEM micrograph of the enamel/steel interface[9]

With SEM, characteristic X-rays emitted by interaction of the electron beam and the sample identify the elements present using electron energy dispersive spectroscopy (EDS). The EDS images of Figure 6 are shown in Figure 7. These confirm the formation of metallic Fe, Co, Cu, and Ni at the enamel/steel interface.

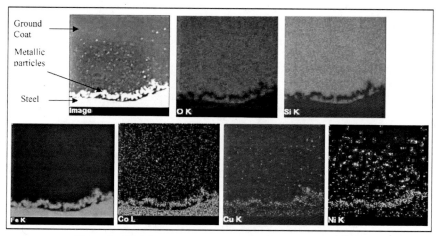

Figure 7. SEM EDS maps of the ground coat/steel interface[10]

The microstructure of enamels contains a bubble structure. In wet-spray enamels, the gases that form the bubbles originate from the thermal degradation of organic material contained in the clays used to suspend the frit particles in water. Otherwise, the bubbles are from gases such as hydrogen, water vapor, carbon monoxide, carbon dioxide and nitrogen emitted by the steel starting below 1200°F (649°C). Excessive bubble can result if the moisture level in the furnace is too high. Excess trapped hydrogen from the steel can cause the spontaneous fracturing phenomena called fishscale.

Andrews observed the evolution of the enamel microstructure during firing.[11] First, the bisque enamel surface cracked, presumably from the expansion of the steel. With increasing temperature, a wavy appearance was observed as the T_g was passed and the enamel became a decreasingly viscous liquid. Then, the enamel smoothed out, and bubbling, which could be violent, began. Finally, large bubbles were eliminated, leaving a fine distribution of smaller ones.

THERMAL EXPANSION

Porcelain enamels develop compressive strength on cooling from the mismatch between the expansion of the substrate and the enamel. The thermal strains that would occur are:

$$\Delta\varepsilon_0 = \int_{T}^{T_0}(\alpha_2 - \alpha_1)dT$$

where α_1 and α_2 are the respective coefficients of thermal expansion for the enamel and steel and dT is the amount of cooling. As seen on **Figure 2**, above T_g, the enamel is in tension because its expansion is greater than the steel. Below T_g, the frit expansion decreases to the value in the linear range, and the tension is relieved by increasing compression.

Figure 8 shows the total mismatch strain from the contribution of the tensile strain above T_g and the linear thermal expansion below T_g. The hard ground coat frit will have the highest residual compressive stress while the soft frit, while potentially good for creating bond, would be at risk from spontaneous delamination from the steel through spalling. Actual results depend on the strength of adhesion, which is determined by the amount of wetting and the amount of cobalt and nickel present in the glass.

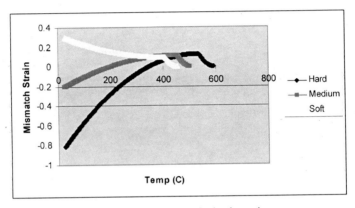

Figure 8. Mismatch strain in the three glasses

SUMMARY

Data was drawn from the literature and laboratory work to illustrate the phenomena that occur while firing a ground coat on steel. Figure 9 shows a typical industrial furnace profile of steel temperature versus time to summarize the points discussed.

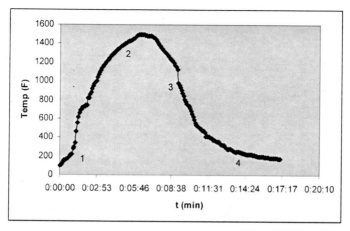

Figure 9. Enamel firing profile

Before firing, the enamel is a dry compact of frit and additive particles. At about point 1 on Figure 9, if clay is present, it begins to dehydrate, and the organic components degrade. This contributes to the formation of the bubble structure. The steel oxidizes, and, as the temperature passes T_g, the enamel becomes a decreasingly viscous supercooled liquid. At point 2, the galvanic reaction, during which the enamel reduces the iron oxide to create adhesion, occurs. While the surface tension of the enamel is a consideration, the viscosity at the firing temperature more strongly determines if the enamel will wet the oxidized steel. After reaching a peak temperature, the part begins to leave the hot zone of the furnace. The enamel cools, becomes more viscous, and contracts faster than the steel. At point 3, the enamel becomes rigid, and the tensile stresses are relieved. If the linear thermal expansion of the enamel is less than the steel, residual compressive stresses begin to build that strengthen at point 4.

Considerable lab work goes into predicting and understanding enamel behavior in service. The fusion flow test gives value of glass viscosity at one temperature compared to a reference standard. Dilatometry provides values for α, T_g, and T_s. These determine if an enamel might spall, how well it might adhere, and when it will soften and begin to "fire out." Lastly, work continues to better understand the high-temperature reactions between enamels and steel that result in bond.

REFERENCES

[1] A. Fluegel, "Glass Viscosity Calculation Based on a Global Statistical Modelling Approach," *Glass Technology: European Journal of Glass Science and Technology Part A*, February 2007, **48** (1), 13-30.

[2] W.D. Callister, *Materials Science and Engineering An Introduction*, Wiley, 1990, p. 434.

[3] R. Ota et al., "High and Medium Range Viscometers and the Their Test with Some Alkali Silicate Glasses," *Journal of the Ceramic Society of Japan, Int. Edition*, **99**, 2 (Feb), pp. 164-168.

[4] C. Baldwin, "Microstructural Engineering of Alumina- and Zirconia-Based Laminates Using Rapid Prototyping," M.S. Thesis. Case Western Reserve University, 1999, p. 7.

[5] S. Lopez-Esteban et al., "Spreading of Viscous Liquids at High Temperature: Silicate Glasses on Molybdenum," *Langmuir*, 2005, **21**, 2438-2446.

[6] K. Sarrazy, "Understanding of Porcelain Enamel Adherence on Steel Improvement of Enameling Process Applying an Interfacial Layer," pH D Dissertation. University of Limoges, 2003, p. 85.

[7] A. Dietzel, "Theory of Adherence of Enamel on Iron," *Ceramic Age*, 1953, pp. 17-39.

[8] B. W. King et al, "Nature of Adherence of Porcelain Enamels to Metal," *J. Am. Ceram. Soc.*, **42**, 1959, 504 – 525.

[9] K. Sarrazy, Op. Cit., p. 60.

[10] K. Sarrazy, Op. Cit., p. 62.

[11] A.I. Andrews, *Porcelain Enamels*, Second Edition, The Garrard Press: Champaign, IL, 1961, pp. 420-421.

STATUS OF PORCELAIN ENAMELING STEEL MARKET

Kim Frey
S&S Porcelain Inc.

I believe in Self Fulfilling Prophecies relative to future market conditions. If you think things will be bad or good...you're right. The glass is either half full or half empty. The published view from the Appliance Magazines 55th Annual Industry Forecast dated August 16th, 2007 gives a fairly benign outlook for the economy in 2007. Companies and economists alike look to 2008 with great anticipation for growth.

Although the United States has lost some industry and strongholds in specific marketplaces, we continue to enjoy a rather robust market. Although new appliance production is certainly impacted by new housing starts, our products and sales are not completely influenced by that measurement alone.

The refinancing and remodeling of existing homes and/or structures are often viewed as favorable as they relate to the impact on products produced from enameling steel. This phenomenon can be realized at all levels of construction, from entry level to high end cooking products, bathroom lavatories and bathtubs to hand dryers.

Unaffected by the matrix of housing starts are other markets and or end use applications for products produced from enameling grades of material. Some of these include:

- Writing Boards
- Signage
- Hot Water Heaters
- Bulk Storage Containment Devices
- BBQ Grills
- Cooking products (i.e. pots, pans, slow roasters, etc.)
- Hospital food warming devices
- Automotive

In summary, business levels in the 4th Qtr 2007 and beyond are encouraging.

GAGE REDUCTION AND THE IMPACT ON PORCELAIN ENAMEL PARTS

Larry Steele
Mapes and Sprowl Steel

ABSTRACT

Why consider reducing the thickness of steel used to manufacture a particular part? Is it because the part weighs too much? Is the weight of the part too heavy for assembly? Why do we want to reduce thickness of the steel used in this part?

Some might call it cost containment. Some might say to save money. Some might say to reduce overall weight........Bottom line, it is..............

!!! $$$$$$$$!!!

Reduced Material Thickness ➔ Reduced Weight per Part ➔ Fewer $ per Part.

Cost savings seem to be on the minds of every manufacturer, regardless of the end product produced. Steels intended for porcelain enameling applications are no exception. However, it is not as simple as just deciding to buy a lighter gage steel product. In the parts that you manufacture, there are many steps in their production and many of these steps will be impacted by "thinner" steel. The purpose of this paper is to highlight some of those areas that must be given consideration. Without proper planning and testing, problems will arise that could end up being disastrous without proper prechange evaluation.

Review Your Manufacturing Process:

- Receive Steel
- Process Steel and Form the desired Part
- Process Part through Cleaning System
- Apply and Fire Porcelain Enamel
- Assemble Part into final product
- Crate and ship your product
- In-service Requirements of the Part in the Final Product

RECEIVING AND STORING THE STEEL

I have always been a big proponent of damage control. This begins at the time of receipt of your steel. It makes no difference if you are purchasing material in blank form or in coil form. It is imperative that steps be taken to insure that your steel is not mechanically damaged prior to being taken to the press area. If a coil is set on a pebble, there will be damage several laps into the coil OD. If a pebble is under the skid on a wood pallet, there will be several blanks damaged on the top of the stack of blanks. When material is taken to the press area, it is mandatory that damage control be at the forefront. Do not set coils directly on the floor; do not stack skids of blanks at the press area without adequate protection between skids. With thinner steel in your coils or blanks, you will damage more material. Damage control is much more important as you enter into gage reduction programs.

FORM THE PART

Once you have moved your damage-free steel to your press area, you will fabricate your parts. What steps are involved in the actual manufacture of your parts? Shearing occurs in nearly every part. With thinner steel, it will be necessary to maintain proper clearances for the shearing operation. This may be a trim operation, it might be a piercing operation, or it might just be punching a mounting, or a screw, hole. Improper die clearances will increase the amount of resulting burr. Excessive burr can cause burn-off. Excessive burr can create assembly issues. Insure that adequate adjustment is available to accommodate a thinner part.

Many parts require bending to produce flanges. Springback is always a design factor when making tooling for a bending operation. Flanges and sections are bent to a specific angle to account for a specific amount of springback after the bending operation. Die clearances will impact the amount of bending actually induced to the part. Again, it is necessary to readjust die clearances to compensate for thinner gage steel. Even with small gage reductions, it will be necessary to 'tighten' clearances in the bending portion of your fabrication operation to insure an adequate amount of 'overforming' when accounting for springback.

Is this a part that is drawn? Remember the Forming Limit Diagram

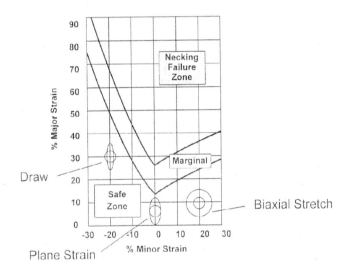

Has Circle Grid Analyses been performed on this drawn part? If so, where do the strains plot? If CGA has not been performed prior to consideration of a gage reduction on a drawn part, it is

imperative that it is done before proceeding!! Are there any strains that are near the marginal zone?

If so, remember to consider the calculation of the FLD_0. FLD_0 is calculated as follows:

$$FLD_0 == [23.3 + 359t] \times (n / 0.21)$$

$$t = \text{thickness} \ \& \ n = n \ \text{Value}$$

It is a safe assumption to believe that the n-Value will not change if the type of product does not change. Therefore, the critical variable is the material thickness. It may be necessary to shim stop blocks in order to accommodate thinner material and control flow into the die. If stop blocks or shut height control is fixed, it may be necessary to modify tooling in order to accommodate lighter gage steel.

Consideration must be given to any braces, lugs, mounting clips, etc. that are welded to the finished part. It is always recommended that these items be of the same material as the part and they should be lighter in gage than the final part. It may be necessary to establish a new inventory item if there is not material currently inventoried that is in the gage range that would be recommended for such attachments with a reduced gage part.

CLEAN THE NEW PART

Consideration must be given to the cleaning system. With a gage reduction, parts will be lighter in weight. The impact of the sprays in the washer can knock parts off hangers. It may be necessary to change the method by which parts are suspended during the cleaning cycle. Likewise, any such changes will need to be evaluated for effects on current hanging configuration relative to other parts going through the cleaner are the same time.

APPLY AND FIRE THE PORCELAIN

If application of the porcelain is via dry powder, electrostatic application, it is unlikely that any changes will be required due to material thickness reductions. However, lighter weight parts going through an automatic flowcoater may require changes in how they are hung in order to minimize chances of being blown off the hangers due to the force of the impact of the enamel slip on the part. Likewise, lighter weight parts will be more easily moved during conventional air spraying of porcelain – either manual or automatic.

Parts must be suspended through the porcelain firing furnace. If parts are being hung by brackets welded to the part, it may be necessary to change hanging method if it is found that there is too much distortion during the firing process. Likewise, many parts are suspended by flanges during the porcelain fire. Although the part will be lighter in weight, with a reduction in material thickness, the flanges may no longer be able to support the weight of the part during the high temperature porcelain fire. It may be necessary to redesign hangers or change the manner in which parts are suspended during the porcelain fire. Enamel warp characteristics must be taken into account, as well. As the porcelain enamel cools, it creates a compressive strain condition in the fired part. Lighter gage steels will be less resistant to the forces created by the enamel warp. This may require formulation changes in order to compensate. However, in-service performance requirements may make changes difficult to achieve. With lighter gages on some parts, there is a possibility of overfiring. This may have a negative effect on overall furnace chain loading and should be carefully evaluated.

It is highly recommended to work with your porcelain enamel supplier in these areas.

ASSEMBLE PRODUCT, PACKAGE, & SHIP

Lighter gage parts will be more susceptible to damage from handling. It is necessary to re-evaluate your handling procedures for finished parts and for assembly of these parts into your final product. Structural integrity of both the parts and the final product can be compromised with thinner steel substrate. Carriers used for transporting finished parts may require changes to insure damage-free parts at the assembly line.

As stated above, lighter gage parts will be more susceptible to handling damage. This is also true in the assembly process. Section strengths may be somewhat compromised by lighter gage steel. This being the case, driving screws may cause more distortion and subsequent cracking and/or chipping of the porcelain enamel. Likewise, handling of the product during the assembly process can cause more flexing than previously experienced with heavier gage parts. It may be necessary modify handling between stages in the assembly process or to modify how subassemblies are transported between assembly stages.

Shipping assembled product to distribution centers or to the final customer will induce strains on the product. These may be vibration strains or they may be impact strains. Side-to-side movement of the final product, as may be experienced during shipping, may flex internal parts to the point that, due to reduced strength from a lighter gage steel substrate, cracking and/or chipping of the porcelain enamel may result. Exposed parts with porcelain enamel must be protected from impact. Reduced steel thicknesses will result in final parts that are less impact resistant. It will be necessary to re-evaluate overall packaging of the final assembled product.

IN-SERVICE REQUIREMENTS OF THE FINAL PRODUCT

Thoroughly examine the in-service requirements of your final, assembled and shipped product. Reduction of steel substrate thickness may require some sort of reinforcement of the finished part in order to withstand loads during use by the consumer. For example, it may be necessary to use a separate formed piece of steel to reinforce a cooktop if the thickness has been reduced to the point that it will not withstand the loads incurred during actual use by the consumer. Insure that all costs associated with a particular part are thoroughly evaluated.

CONCLUSION

A number of areas for consideration have been listed and discussed. Not all of these will apply to every part considered for a reduction in the steel substrate thickness. However, whenever steel thickness is being evaluated for a cost savings program, it is imperative that many areas in the overall manufacturing process of making and shipping your final product be considered. It is not that difficult to purchase a lighter gage steel substrate. It is not that easy to answer all of the questions that center around steel thickness as affects your final assembled product for the consumer. Consider the following list of questions whenever any part is being considered for a gage reduction program:

- How good are your damage control efforts in your steel receiving, storage, and press areas??
- Are you able to adjust your die clearances tighter to insure that you maintain minimum burr in areas where the steel is being cut/sheared??
- Is there adequate adjustment in your tooling to accommodate bending of lighter gage steel?
- Is the part drawn? If so, where do strains plot on FLD? Recalculate FLD_0 to insure that strains do not become marginal or critical.
- How is shut height controlled? How are stop blocks currently set up? Are there shims that can be changed or are the stop blocks fixed and, possibly, require tooling modifications to accommodate reduced material thickness?
- Is material available from which lugs, brackets, mounting clips, or other attachments can be fabricated? Will a reduction in part thickness necessitate that a new inventory item be created to manufacture attachments from the proper thickness in order to avoid problems in porcelain enameling? Should these attachments be made from a thinner material, will they still satisfy the requirements for which they are intended?
- Will material thickness reductions necessitate special hangers in the washer? Will any required changes be compatible with the current configuration of your washer hanger system or will any changes require that the hanging system be significantly modified? If the answer is YES, what effect will this have on hanging of current, unaffected parts in your washer?
- What changes will be required in the part hanging system for the application of the porcelain enamel? Will there be any negative impact on other parts going through the application system at the same time?
- What changes will be required in hanging of parts on the furnace chain? If changes are required, what impact will there be on other parts going through the furnace at the same time? Will porcelain enamel warp characteristics be detrimental to part integrity with a gage reduction? Is there a possibility of overfiring? What will be the effect on overall loading of your furnace?
- What changes will be required in moving the porcelain enameled parts to the assembly area? Will it be necessary to modify part/subassembly carriers?
- What changes will be required to your assembly operations as the product progresses through the assembly process?
- Does your current packaging offer sufficient protection of the product when using lighter gage steel for various parts?
- What additions and/or changes will be required in order for the final assembled product to meet in-service consumer requirements?

CAST IRON PROCESS CONTROL IN A DEVELOPING MARKET

Liam O'Byrne
HKF Industries, Inc.

INTRODUCTION
This paper will share some key process control parameters that need to be addressed when sourcing cast iron product from the Chinese market for the Porcelain Enamel Industry. Statistics presented in the paper has been obtained from various sources and personal experience based on working in the Chinese market over the last three years.

MARKET SITUATION
According to documents researched on the Chinese foundry industry, there are somewhere between 12,000 – 13,000 foundries currently operating in China. Of this number, 100-200 are considered to be large and/or modern foundries in terms of their technology and market presence, while the remainder varies in size from mid-size to very small.

In such a market situation, it can be difficult to accurately determine a lot of up to date information on the status of many of the smaller foundries in the market and their level of technical knowledge and production controls. The Chinese government is addressing this issue by significantly improving the amount and quality of third-level education opportunities in the foundry specific technologies.

A recent survey of the China Foundry Association web site listed 35 foundries as making iron castings. While the Association obviously promotes member companies, the fact remains that immediate identification of a suitable Chinese foundry for enamel quality castings is not always as easy as might be initially thought, and this is generally the biggest challenge when beginning the search for a suitable supplier.

BUSINESS RELATIONSHIPS
In addition to finding a suitable casting manufacturer, the business relationship with the Chinese company must be carefully considered. In Chinese business, the concept of Guān Xi (关系) plays a very important part. Loosely translated, Guān Xi means connections or relationships. It is a key component of many Chinese business operations and is a system that is based as much on mutual trust and understanding as on formal contractual responsibilities. Partnerships in business rely on personal relationships between key individuals in the companies doing business together.
It has been said that with good Guān Xi, difficult business issues are often very easily resolved, while without Guān Xi, seemingly simple problems can prove to be impossible.

FOUNDRY CAPABILITIES
Assuming that initial contacts with a suitable foundry have been made, what are the main concerns for a company beginning to do business with a Chinese foundry?
Key items to be considered and discussed in this presentation are;
- Access to and control of raw materials.
- Control of In-house foundry processes.
- Obtaining accurate and timely information from the foundry.

RAW MATERIALS

The main items involved in raw material supplies are generally the products used for melting the iron, and the materials used in the chemical composition of the iron. With regard to iron melting, electrical melting is fairly self-explanatory, but in the more common (in the Chinese industry) Cupola melting, the availability and quality of Coke for melting purposes is a key consideration. Coke has both physical and chemical requirements for ensuring the best melting process in a cupola. Two main physical components of coke quality are its Stability and CSR (Coke Strength after Reaction).

Stability is defined as the ability of the coke to withstand breakage at room temperature, which is important when ordering a specific average size of coke material. The foundry will want to be sure that the coke that is delivered is not significantly reduced in size during the handling and transportation from the coke manufacturing site.

CSR is the ability of the coke to break into smaller pieces under high temperature reactions in the melting furnace. This can be an advantage to helping achieve appropriate Carbon pickup from the coke as the metal percolates down through the cupola coke bed by increasing the surface area available for reaction.

The key chemical components of the coke are the amount of Fixed Carbon, Moisture, Ash, Sulfur, Phosphorus and Alkalis present in its composition. In this list, all items except the fixed carbon are considered undesirable and should be kept to a minimum. It is advantageous to know if the foundry specifies the chemical and physical requirements for their coke material and how they check and control it.

Pig Iron

The key outside source of controlling iron chemical composition is pig iron. Generally a certificate of composition is provided with each delivery, and it is important to determine if the foundry has a specified chemical composition for its pig iron, and is it met by their customer? In addition, does the foundry regularly check the pig iron supplied or do they just accept whatever is delivered?

In-House Foundry Controls

The foundry must also have sufficient control of their own molding and casting processes to assure the customer of consistent quality. These main controls include Sand System and Metal Melting Controls.

Sand System Control

While more foundries are using automatic molding equipment for the preparation of sand molds, there are still quite a large number of smaller and mid-size foundries using manual or floor-molding operations to prepare the sand molds for iron castings. No matter which method is used however, the sand preparation for the molds needs to be controlled.

Key parameters that any sand system control needs to address are:

- Green Strength
- Compactibility
- Moisture Content
- Active Clay Content

These parameters should be measured regularly and be demonstrably kept between the approved tolerances set by the foundry processing manuals.

The metal melting controls should at a minimum include metal composition analysis on a regular basis and pouring temperature measurement and control. There are modern, accurate and very quick ways of doing this on an ongoing basis, and more Chinese foundries are implementing these controls on a consistent basis.

ONGOING RELATIONSHIPS

Finally, once a suitable casting supplier has been found and the business relationship begins, it is important to build a solid system of regular contact with the foundry in the spirit of improving and developing Guān Xi.

Regular visits to the casting supplier should be implemented and a continuous improvement philosophy should be shared on both sides of the relationship. It is a well stated fact, that good suppliers are generally developed by good customers who clearly and regularly share the expectations they have for the product, and support the supplier in any way possible in meeting those expectations.

It is also a personal experience that changes in the processes that are required from time to time are better implemented deliberately and carefully, taking perhaps a little more time than might generally be expected with other suppliers, in order to make sure that the requirements for the change and the potential benefits are well understood by all parties beforehand. Once the benefits are clearly explained and understood, ownership and successful implementation are much easier to obtain and keep in place during future production.

RECYCLE-FRIENDLY AQUEOUS CLEANERS

Elizabeth J. Siebert, William G. Kozak, and William E. Fristad
Henkel Corporation

ABSTRACT

Metal cleaners must have a number of desirable properties to be technically effective and successful in the marketplace. They must, obviously, remove a wide variety of soils, prevent the redeposition of the suspended soils, and continue to clean well even when highly contaminated. Additionally, they must have appropriate foam control, rinse off from the part surface easily, and operate at an acceptable overall process cost. The real measure of a successful metal cleaner is that it prepares the surface for the next step in the process of assembling a final product. The next step is typically the application of a thin layer such as paint or porcelain enamel.

MECHANISM OF CLEANING

The overall operational steps in the cleaning process are illustrated in Figure 1 and involve wetting the surface, soil displacement and removal, and soil encapsulation to prevent redeposition. A number of factors reduce or extend the lifetime of a cleaner bath by affecting one of these three steps. Three main factors, along with drag out, reduce cleaner bath life through contamination with oil, grease, metal fines and other particulates. (1) Emulsified oil effectively ties up the surfactant and reduces further cleaning effectiveness. (2) Metal ions build up in the bath due to hard water evaporation or removal from the part surface, and the chelating or complexing ingredients are rendered ineffective for further complexation. (3) Atmospheric carbon dioxide reacts with the caustic, sodium or potassium hydroxide, in an alkaline cleaner and converts it to carbonate ion. This is a natural reaction in any alkaline cleaner and occurs particularly fast in a spray cleaner, where the fine spray particles create a large surface area for reaction with carbon dioxide.

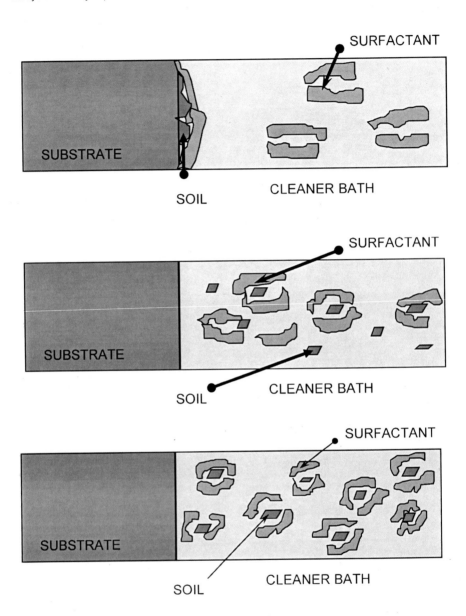

Figure 1. Steps in Soil Removal

FACTORS AFFECTING CLEANER RECYCLE

The two factors that can extend cleaner bath life before dumping and recharging are (1) continuous or periodic replenishment of the components as they are "consumed" as described above or removed from the cleaner bath by drag-out, and (2) removal of the oils and particulates that build up in the cleaner by filtration of particulates and separation of the oils by either an oil-water coalescer or ultrafiltration. Particle filtration is well known and will not be further discussed here. Oil separation by use of either an oil-water coalescer or ultrafiltration is also known, and this paper will not focus on the details of the use of the devices, but more on how the use of these devices require careful control of the cleaner bath chemistry if it is to remain effective.

An oil-water (O/W) coalescer allows the oily, floating phase to be removed in an oil-splitting cleaner. i.e. a cleaner formulated to not strongly emulsify oil once removed from the surface. When agitation is reduced, as in an O/W-coalescer, the oil phase splits out readily and can be removed. However, this oil phase may contain dissolved surfactant, and, therefore, surfactant is lost with the discarded oil phase. Table 1 shows oil partitioning of typical surfactants used in alkaline cleaners into a mineral oil phase. It becomes obvious that an alkaline cleaner relying on typical nonionic surfactants, as exemplified by the first four entries in Table 1, will rapidly lose their cleaning power as the surfactant is depleted. This loss of surfactant must be compensated for in a replenishment program.

Table 1. Surfactant Solubility (Partitioning) in an Oil-Containing Cleaner

Partitioning of Surfactants in 2 Wt% Cleaner With 5% Mineral Oil at 130° F
(Samples Stirred for 1 Hour and Allowed to Separate for 24 Hours)

Surfactant	Type	Chemical	HLB No.	Nonionic Surfactant Titration		% Surf Left in Aq. Phase
				Initial	Final	
Pluronic L-61	Nonionic	Polyoxypropylene-polyethylene Block Copolymer	3-7	2.8	1.8	64%
Triton DF-12	Nonionic	Modified Polyethoxylated Alcohol	10.6	2.7	2.2	81%
Tergitol NP-9	Nonionic	Nonylphenol ethoxylate (9 moles of EO)	12.9	3.5	2.2	63%
Igepal CO-850	Nonionic	Nonylphenol ethoxylate (20 moles EO)	16	3.7	2.8	76%
Proprietary Surfact. #1	--	Confidential	--	27	27	~100%
Proprietary Surfact. #2	--	Confidential	--	25	27	~100%

In ultrafiltration, surfactants are lost not only due to solubility in the oil phase itself as shown in Table 1, but also by the process of oil emulsification. By either process, surfactant is disposed of with the rejected oil phase (retentate). Ultrafiltration is more commonly used with oil-emulsifying cleaners, i.e. cleaners formulated to keep the oily soils in suspension as a fine emulsion, rather than letting them separate and float to the top of the bath. To make the oil emulsion particles stable in the cleaner bath, a sheath of surfactant molecules is required to coat each oil droplet. This means that as the oil droplet, or oil emulsion particle, is rejected by the ultrafiltration (UF) membrane, a portion of that oil particle is actually surfactant. Thus, theoretically some surfactant must be lost with each emulsified oil droplet.

However, even surfactant that is not emulsifying oil can be lost by ultrafiltration. The pore size of an ultrafilter membrane (0.001-0.5 micron) is of the same order of magnitude or smaller than the surfactant micelles in the cleaner bath. Surfactants are not normally truly in solution, like dissolved salts would be, but exist in small (approx. 0.1 micron) aggregates called micelles. The existence of surfactant micelles is seen in Table 2 where the same surfactants used in Table 1 were put into an alkaline cleaner formulation, and this unused bath subjected to ultrafiltration with a nominal pore size of 0.4 micron. Table 2 shows that at a typical bath operating temperature (140° F) often more than 50% of the surfactant is lost on every pass through the UF membrane. Since the UF process is continuous, eventually many passes are made and all the surfactant would be removed and discarded in the retentate.

Table 2. Surfactant Removal by Ultrafiltration

Replacement Surfactant				Permeate Rates (ml/min)		%Surfactant Passing through Membrane (30 psi)	
Name	Type	Chemical	HLB #	RT	140 °F	RT	140 °F
Pluronic L-61	Non-ionic	Polyoxypropylene-Polyethylene Block Copolymer	3-7	54	95	95%	25%
Triton DF-12	Non-ionic	Modified Polyethoxylated Alcohol	10.6	62	ND	95%	30%
Tergitol NP-9	Non-ionic	Nonylphenol ethoxylate (9 moles of EO)	12.9	53	100	70%	35%
Igepal CO-850	Non-ionic	Nonylphenol ethoxylate (20 moles of EO)	16	61	107	95%	90%
Proprietary Surfactant #1	--	Proprietary	--	52	86	~100%	~100%
Proprietary Surfactant #2	--	Proprietary	--	54	96	~100%	~100%

ULTRAFILTRATION TESTING

A series of experiments was conducted in order to determine the effect of ultrafiltration on actual alkaline cleaner formulations. Figure 2 shows the overall schematic of the experimental UF system used to generate the data reported below, which is similar to many commercial UF units. A portion of the used alkaline cleaner is transferred to a smaller retentate recycle tank. From there the UF membrane is fed by a medium pressure pump with only a small amount of the cleaner feed actually permeating the UF membrane and being returned to the cleaner tank. The majority of the cleaner feed is retained by the membrane and returned to the retentate recycle tank. This use of cross-flow filtration prevents the rapid plugging of the UF pores and allows more efficient operation. By continuing to pump the recycle retentate tank contents past the UF membrane, the oil phase is allowed to build up in the retentate recycle tank before disposing it. The oil-enriched retentate recycle tank is discharged when the permeate flux rate has decreased to a predefined level. The UF membrane is then cleaned and ready for the next cycle.

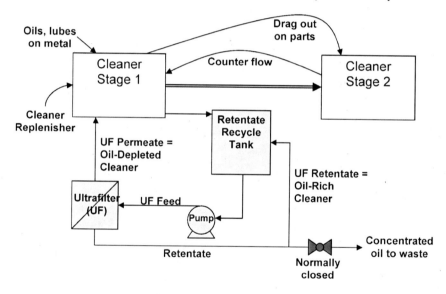

Figure 2. Ultrafiltration Schematic Diagram

The net result of ultrafiltration is that oil is removed from the alkaline cleaner bath. However, the two real questions are what else besides oil is removed and what remains behind in the cleaner? If surfactant is removed with the oil, then it will have to be replaced. If other contaminants pass through the UF membrane and are returned to the cleaner tank, how will they affect the cleaner bath life?

RESULTS

A typical commercial alkaline cleaner based on nonionic surfactants was tested according to the UF scheme shown in Figure 2 and the results are summarized in Figure 3. In this experiment, the cleaner bath was loaded with 0.5% emulsified oil, and the contents of the retentate recycle tank subjected to one complete pass across the UF membrane (one turnover). It is readily seen that there is no oil in the permeate (squares), which is returned to the main cleaner tank, while the oil in the retentate recycle tank (triangles) increases, since it is rejected by the UF membrane. Thus, oil removal by ultrafiltration occurs as desired.

Fate of Oil in Cleaner Testing

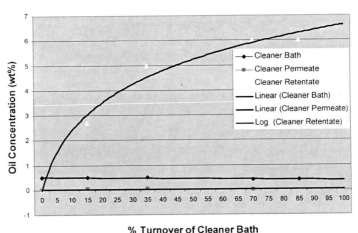

Figure 3. Fate of Oil in Typical Alkaline Cleaner

Figure 4 shows the fate of the nonionic surfactant in this same alkaline cleaner. The nonionic surfactant behaves much like the oil did in Figure 3. The concentration of the surfactant in the permeate (squares) is only about 20% that of the starting cleaner bath (diamonds). Thus, as the UF operation proceeds, the surfactant continues to be depleted in the cleaner bath and builds up in the retentate (triangles) where it would eventually be disposed of along with the oil.

A special recycle-friendly cleaner was formulated using the proprietary surfactants listed at the bottom of Table 1 and 2. These surfactants are not soluble in the oil phase and do permeate through the UF membrane very efficiently. In Figure 5 it is readily apparent that the oil is still retained by the UF membrane and builds up in the recycle retentate (triangles), while the permeate contains no oil (squares). However, the important difference is seen in Figure 6 where the concentration of surfactants remains at 100% of the original level in the cleaner bath (diamonds) and in the UF permeate (squares), while no surfactant is held up in the retentate (triangles). Such a cleaner would be much more efficient with UF oil removal because the surfactant is not lost with the disposed oil.

Fate of Nonionic Surfactant in Cleaner Testing

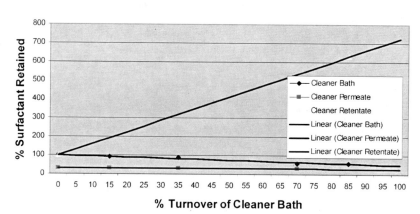

Figure 4. Fate of Nonionic Surfactant in Typical Alkaline Cleaner

Fate of Oil in Testing of Recycle-Friendly Cleaner

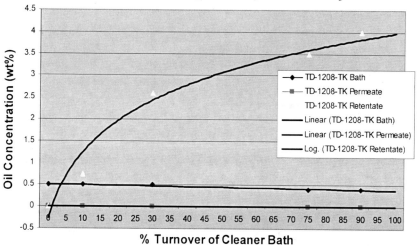

Figure 5. Fate of Oil in Recycle-Friendly Alkaline Cleaner

Fate of Surfactant in Testing of Recycle-Friendly Cleaner

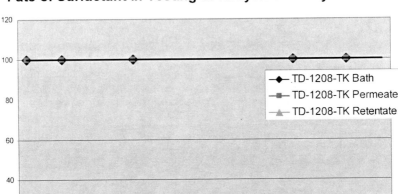

Figure 6. Fate of Surfactant in Recycle-Friendly Alkaline Cleaner

Thus, it is possible to design an alkaline cleaner so that the surfactants are not lost with the disposed oil; however, most commercial alkaline cleaners today will lose surfactant with the oil.

The other major component of any alkaline cleaner is the alkalinity. Strongly alkaline cleaners contain sodium or potassium hydroxide, and their cleaning ability is generally directly dependent on the amount of hydroxide ion present. Hydroxide ion is often referred to as free alkalinity by an industry standard titration. The hydroxide can be lost by reaction with fatty acids that may be present in some lubricants or other acidic soils on the metal surface being cleaned. However, the major loss of hydroxide ion is in a reaction with carbon dioxide in the air to form carbonate ion. Carbonate is not nearly as effective in cleaning as hydroxide ion. The sum of the carbonate + hydroxide ion is referred to as total alkalinity. The rate at which an alkaline cleaner reacts with the air is dependent on several factors, but primarily it needs contact with air, i.e. spray reacts faster than immersion.

Figure 7 summarizes the important changes in carbonation of a spray alkaline cleaner bath. In this experiment a small 5-gallon cleaner bath was sprayed without exposure to any soils. The pH of the cleaner dropped from 13 to 10 as the hydroxide ion reacted with carbon dioxide to generate a carbonate buffered cleaner solution. This drop in pH was essentially independent of temperature over the range of 100-140° F. Much more dramatic is the drop in free alkalinity (FA) or hydroxide concentration. In this cleaner it dropped from 9.5 points titration to 3.5-4.0 over the 5-day experiment. During this same time frame the total alkalinity (TA) shows very little change, as expected, because the sum of the carbonate + hydroxide remains constant at approximately 12 points.

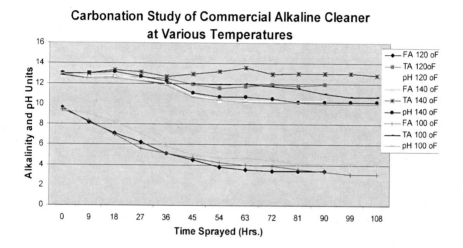

Figure 7. Carbonation of Alkaline Cleaners

CONCLUSIONS

Ultrafiltration of most traditional alkaline cleaners will remove surfactant along with the oil and require that the surfactant be replenished regularly. However, monitoring the actual concentration of surfactant in a used cleaner is difficult without sophisticated analytical equipment. So it is difficult to run an efficient cleaning operation. Alkalinity is lost due to reaction with carbon dioxide in the air, especially rapidly with spray cleaners, and forms carbonate ion which is much less effective in cleaning. Carbonate ion passes through the UF membrane and, therefore, builds up in the cleaner bath. To control the ratio of free alkalinity (caustic) to total alkalinity (caustic + carbonate), the cleaner bath still must be regularly decanted or overflowed to eliminate carbonate ion. Thus, ultrafiltration or O/W-coalescers can extend alkaline cleaner bath lifetimes, but either special maintenance or cleaner formulas are required.

COST DRIVEN DEVELOPMENT OF NEW CLEANERS

Kenneth R. Kaluzny
Coral Chemical Company
Waukegan, IL

ABSTRACT

Now more than ever we need to operate as efficiently as possible to compete with foreign manufacturing. This presentation will address cleaner chemical strategies that can reduce operating costs related to your finishing system.

I have identified four current approaches used to reduce cleaning costs. There are more ways to produce operational savings but only four are covered under the scope of this paper. The four chemical approaches to operational savings include making the cleaner formulation less robust, making the cleaner operate at a lower temperature, making the cleaner last longer reducing dumping and recharge costs, and making the metal working lubricants easier to remove and less damaging to the cleaning operation.

OVER FORMULATED CLEANER CHEMISTRY

One way to reduce cleaning costs is to make the cleaner less robust. Why buy a Cadillac when a Chevy can get you where you want to go? This approach doesn't mean dilute your cleaner formulation with water. As much as this would make a cleaner less expensive per gallon it doesn't decrease the cleaner's operating cost. Diluting the product will unnecessarily increase your freight costs as well as handling costs. This approach assumes that the current cleaner is over formulated for the soils involved. You might have a cleaner that is built to remove difficult organic and inorganic soils and your quality requirements may not require this level of cleaning. For example, product cost reduction refers to minimizing builder expense by using lower cost builders such as hydroxide and ash rather than poly or ortho-phosphates. The cleaner cost can also be reduced by minimizing or using lower cost detergents. Dispersants or anti-redeposition agents used to lengthen bath life may not be necessary. Identify your soils and perform loading studies to determine what makes sense for your operation.

ENERGY SAVING CLEANER CHEMISTRY

Lowering the cleaner's operating temperature is a potential manner in which to lower your cleaner operating cost. Typically there is no reason to change the builder system. Common builders for those who aren't familiar with cleaner formulations include phosphates, silicates, hydroxides, borates, and carbonates. For spray cleaners the surfactants will definitely have to change. Typically nonionic surfactants are used in spray cleaner formulations as they are typically lower foaming detergents than cationic and anionic detergents or soaps. However, as you lower the operating temperature even the nonionic surfactants foam more. The obvious benefit is reduced energy consumption. Scale and sludge formation will also be reduced because there will be less water added to the tank and less evaporation to condense the solids in the water. Thus there will be less water solids to form sludge and scale. The biggest concern I have for lowering the operating temperature is cleaning. Typically the lower temperature surfactants are better at lowering surface tension than behaving as a detergent. Some lubricants have high pour points, the temperature when they become fluid, requiring temperature to make them fluid to facilitate removal. Otherwise they are very stubborn and hard to remove. Removal of difficult soils can be enhanced with solvents. Solvency may or may not be a concern. Solvents can help to dissolve these stubborn soils. However solvents may cause environmental issues and reporting issues if VOC and FOG limitations are an issue with your operation.

CLEANER LONGEVITY

Another way to decrease cleaner costs is prolong dumping and recharging costs. This may initially seem in contradiction to a previous comment, but it is another strategy. The appropriate chemical strategy will relate to an individual finisher's situation. If the metal finisher is dumping every week, then increasing the cleaner's chemistry would be wise. This would reduce recharging costs. If a user dumps their tank once/week, then think of the cost savings from extending bath life to two weeks. You basically reduce the costs associated with 25 dumps and recharges. This also reduces freight, disposal, and labor costs associated with dumping and recharging the cleaner tank. If applicable, this would also reduce water treatment or hauling charges.

METAL WORKING/METAL FINISHING COMPATIBILITY

Another option to reduce cleaning costs is to enhance the compatibility of the metal working fluid and the cleaner. In short you want to make the forming lubricant easier to remove with the cleaner. Ideally the lubricant would be formulated to cause as little negative impact on the cleaner as possible as well as being easy to remove. The lubricant can also be formulated with agents that act as a replenishing agent for the cleaner. Specifically we are talking about surfactants as they can provide slip for metal working needs and detergency for cleaning. The right lubricant chemistry can also help to reject oil in the cleaner stage which will decrease cleaning costs.

CONCLUSION

How to proceed with the union of a lubricant and cleaner? You need to define your goal; what is it that you want to improve. You may want to increase productivity or improve your efficiencies. Overall you want to reduce your product's cost. The hardest element of the lubricant/cleaner union is the reformulation of products by your vendor. Your work will involve taking the time to perform the metal working and cleaner testing. What you need to do is start by developing a good idea. So take action to help keep your business running efficiently. It will take some work but most if not all things worth anything require work. The work will also result in your professional and personal fulfillment. Partner with your vendor to develop a plan to make your finishing operation more competitive with foreign manufacturing.

2C/1F ENAMELLING PROCESS - A GROWING DEMAND

Hans-Juergen Thiele
EIC Group GmbH member of DEV
Germany

INTRODUCTION

The purpose of this paper is to establish a review of various 2 C/1F systems, which all contribute to an environmentally favorable application system along with energy savings, increased production, capacity and efficiency. In the industrialized countries there is an increasing demand for enamel application systems that will reduce on going environmental issues. Addressing these concerns will help to bring pollution under control and possibly, to be brought in the future to a standstill. To realize the goal of halting pollution, more and more the principle of "the one who pollutes should pay" will apply. Organizations will be forced to take more stringent, compliant measures. The driving forces not only concentrate on the environment, but also on the surface quality, economic use of enamel and automatic application flexibility. Along with the use of low carbon steel with 0.03 – 0.08% C instead of the more expensive decarburised steel quality with 0,003 – 0,008% C A positive contribution to the environment and cost saving can be achieved by:

a. The known process of pre-treatment of the sheet steel by degreasing and rinsing rather than the acid pickling and the nickel treatment
b. Increased enamel material transfer efficiency by utilizing spraying systems with high material transfer efficiency
c. introduction of low fluorine and fluorine free enamels
d. The re-use of recovered enamel as much as possible
e. Spray booths concept designed as closed systems. The air is cleaned from overspray material and carried back into the working environment by passing through a dry filter
f. The re-use of water by using a waste water cleaning and water recycling system.

REVIEW OF THE 2 C/1F ENAMELLING SYSTEMS

The following 2 C/1F systems are in use today:
comparison between the 3 processes:

a. 2 C/1 F – Powder / Powder
b. 2 C/1 F – Wet / Powder
c. 2 C/1 F - Wet / Wet

In principle, these three systems are based on so-called liberty ground coat enamel combinations and therefore suitable on sheet steel, which is just degreased and not acid pickled.

The application technology is based on the following systems:

a. Powder-Powder
The powder ground coat and the powder cover coat enamel are applied on top of each other by electrostatic powder application. No drying is needed. The coating is then fired simultaneously.

b. Wet-Powder
The thin ground coat is applied in the wet system by means of electrostatic or robot spraying and then completely dried. After a very short drying process with infra-red drier the powder cover coat enamel is applied by electrostatic application and then fired.

c. Wet-Wet
The thin wet ground coat is applied by means of electrostatic application or dipping (also possibly by electrophoresis). As no drying is required after this process, the cover coat is applied wet by electrostatic means. After complete drying the enamel layers are fired together.

THE STEEL QUALITY AND THE STEEL PRE-TREATMENT

- Although frequently low carbon steel is also used in the application of the powder-powder system a decarburized steel quality is recommended, in particular when the steel is coated on two sides.
- Until now for the wet-powder system (de-greased) decarburized steel was used. Recently acceptable results (one side application) have been reported in using normal steel quality.
- With the 2 C/1 F wet-wet application system good results have been obtained on low carbon steel.

Generally speaking, we can say that the requirements of the customer in relation to the surface aspect might contribute to the choice of the steel quality. As far as the pre-treatment of the steel is concerned the following points should be taken into consideration.

- The sheet steel should be well de-greased to prevent undesired gas reactions during the firing cycle
- Because in the three mentioned 2 C/1F systems a liberty ground coat enamel combination is applied the acid pickling of the steel is not necessary. Pickling should therefore be only used when corrosion remains on the steel surface.

THE VARIOUS ENAMEL COMBINATIONS FOR THE 2C/1F SYSTEM

A. The ground coat

The purposes are:
- to achieve a good enamel adherence by using sheet steel, which is only de-greased
- to get a faster out gassing from the ground coat before the cover coat is closed
- to avoid black spots where we recommend to use a lower softening temperature and higher surface tension

B. The coating thickness of the ground coat enamel

In principle the ground coat layer has to be thin but thick enough to allow various repairs and firing cycles. (30-40 microns)
- A thicker ground coat layer means more gas bubbles, which can pose the problem of so-called pinholes in the cover coat enamel.
- A thick powder ground coat enamel may lead to more heavy gas reactions due to organic mill additions to the powder enamel.

A wet ground coat requires an electrostatic milling fineness, to obtain a tight layer on the cover coat enamel (3 – 5 / 16900 mesh). Slip parameters for wet ground are:
- specific weight: 1,45 – 1,60 g/ccm
- density: 3-4 g/dm²

For the 2C/1F process a very thin layer on all surface areas is always required. To assure a thin ground thickness (depends on the frit suppliers who have different requirements between 20 – 60 micro m) a fully automatically applied coating is needed. The application system for both ground at wet/powder and wet/wet is the same.

C. The cover coat enamels

It is recommended to choose a cover coat enamel which remains "open" on the surface during the firing cycle and fuses at a later time. In this way, bubbles, which have possibly already risen from the ground coat, can escape. In practice, this means using cover coat enamels with a higher softening temperature.

Slip parameters for wet cover coat are:
- specific weight: 1,55 – 1,65 g/ccm
- density: 12-14 g/dm²

FIRING CONDITIONS FOR 2 C/1F 1 FIRE SYSTEM

Generally speaking there are no special firing conditions needed compared to 1C/1F or 2C/2F. Except for powder where the air curtains may have to be adjusted.

THE 2 C/1F SYSTEM COMPARED TO THE DIRECT-ON WHITE ENAMELLING SYSTEM

A major advantage of the 2 C/1F system is the possibility of replacing the direct-on white enamelling. The direct-on enamelling was introduced in the early sixties and presented extraordinary advantages to the traditional 2 C/2F enamelling system.

The advantages, of course, were that titanium enamel could be applied directly on decarburized steel with a thin nickel layer, just one enamel layer and one firing. A disadvantage of the very complicated steel pre-treatment (intensive acid pickling and the nickel treatment) were less important than the big advantages like energy saving, enamel savings and the increase in the production capacity. However, today the expenses for cleaning the waste water are a burden on the budget (pre-treatment baths). It can therefore be said that the trend is to invest into the 2 C/1F application system instead of the direct-on white enamelling. This choice also seems to be justified from a technical point of view, certainly with regard to the results obtained today.

Finally, it can be said that when applying a 2 C/1F system the steel pre-treatment can be integrated into the enamel application process, leading to less handling of the item to be enamelled. The complicated charging and discharging of the pre-treatment installation is therefore something belonging to the past. In new installations, passivation can be eliminated as the parts go to the pre-treatment through the drier and directly to the various application systems.

The Main Advantages Of The 2C/1F System Powder-Powder
- large production quantities of flatware
- clean system a very high degree of automation
- no drying
- no milling
- possible smooth surface of the cover coat

Disadvantages of 2C/1F Powder-Powder
- difficult with complex design
- necessity of air-conditioning
- limited choice of colors of the cover coats
- high investment cost
- constant recycling of big quantity of powder mixed with fresh powder
- very high abrasion / wear at powder spraying nozzles

Main Advantages of the 2 C/1F Wet-Powder System
- a coating thickness up to 80 microns is possible
- large production quantities of flatware
- clean system a very high degree of automation
- higher flexibility for items with a more complex design

Disadvantages of 2C/1F Wet-Powder
- limited choice of colors of the powder cover coats
- necessity of air-conditioning

The Main Advantages for the 2C/1F Wet on Wet
- quick color changes within less than one hour
- saves energy for firing only once, less oven furnace capacity necessary.
- saves energy for one drier; no drier between the 2 coats
- shorter through-put time

Disadvantages 2C/1F Wet/Wet
- too thick ground coat can cause defects on the surface of cover coat:
  possible capillary action causing grey lines
- too thick ground coat can cause defects on the surface of cover coat.

2 C/1F System Compared with a 2 C/2 F System :
- reduction of working time
- less enamel consumption (thinner ground layer)
- energy-saving (firing just once instead of twice)
- high degree of automation
- larger production capacity
- smaller floor space needed
- lower cost per enamelled item

Alternative to 2C/1F Wet-Wet: 3C/2F
As an alternative to the 2 C/1F wet-wet system, the 3 C/2F wet-wet system used mostly at bath tubs and architectural manufacturing companies. This system requires a thin layer of ground coat enamel applied on the steel and on top of this, wet on wet, a thin layer of cover coat enamel. This first layer of cover coat enamel may come from recovered or waste enamel and may be slightly contaminated. After drying and firing the second cover coat layer will be applied. This second coating does not contain any waste enamel and the second layer is not contaminated. After drying the enamel is fired again.

SUMMARY

That completes my description of the process 2C/1F. So, to summarize we can say the following:

By choosing and determining a new system technique, the most important criteria are modifications to the production system without having high and costly new investment
- high grade of automation
- reduced energy costs
- saving labour costs
- reduction on waste material
- flexibility in enamel adjustment;

And, last but not least,

- a reduction of the ecological damage

Although it is known that a better surface appearance can be obtained in the dry/dry process by using decarburized steel, it has to be said that the negative aspects of this system can be sensitive on the final surface appearance in relation to variations in the ground coat thickness, for example the edges of the product to be enamelled. I am sure that above-mentioned advantages will induce a lot of manufacturers to consider favorably the 2C/1F wet/powder or wet/wet process.

POWDER COATING COLORS: PORCELAIN ENAMEL VERSUS PAINT

Holger Evele
Ferro Corp.

The forces that act upon each individual powder particle - air resistance, weight, aerodynamic force, and electric force – are the same for both paint and porcelain enamel. The impact of each does vary however. The specific gravity of powder paint is close to 1.5 while powder porcelain is closer to 2.5 to 2.7. This added mass increase the impact of forces such a gravity and decreases the ability of the meager electrostatic attraction overcome particle inertia to remove the powder form the air stream during spraying and keep it adherent to the substrate.

The following five factors impact the spray performance of electrostatic applied powder. With powder porcelain, most have greater variation than powder paint.

- Powder Formulations
- Powder particle size distribution
- Powder storage
- Powder age
- Powder recirculation

The diagram below describes the process commonly used to manufacture powder paint. The same flow line exists for powder porcelain. In the case of porcelain the temperature need to be much higher and materials melt and chemical combine with the base glass altering the composition. Due to the complex reactions that can occur at these temperatures, solvation (solution chemistry changes) crystallization, reduction and oxidation differences the end products color becomes more difficult to predict requiring extra quality assurance and increasing costs as less usable material is produced and increasing overall processing expenses.

Making Powder Coating Paint

Thermosetting powder coatings consist of pigments and additives dispersed in a film forming binder of resin and curing agent, milled into a fine powder. There are eight stages in the manufacture of powder coatings.
Weighing of raw materials
Premixing
Extrusion
Cooling and kibbling (breaking of the extrudate into small chips)
Grinding and classification (splitting of the powder into two size fractions)
Sieving
Homogenizing or dry blending (if necessary or specified)
Packing
Of these the key operations are **premixing, extrusion, grinding and sieving**

Premixing
The purpose of premixing is to prepare a homogeneous mixture of the raw materials prior to extrusion according to the product formulation. It is at this stage that any colour-matching adjustments are made.

The David Howell Consultancy

Powder Coating Paint Line

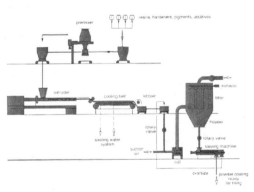

Flow chart for coating powder manufacture (with direct filter box collection)

The David Howell Consultancy

Since the glass takes everything into solution and then may crystallize, some materials with the same chemistry smelted under minor different parameters can give significantly different colors as outlined below.

Contributing To Color

- **Transparency and gloss of coating**
- **Addition of opacifiers**
  - **Titanium Dioxide**
  - **Zirconium Dioxide**
  - **Calcium Carbonate**

- **Dissolved Chemistry of Metal Ions**
- **Ceramic Mixed Metal Oxides [Pigments]**
- **Crystallization of Various Oxides**

Since the chemistry that can be potentially used is more limited by the higher operating temperatures and therefore manufacturing of porcelain enamel versus. paint may result in differing colors in differing light sources.

Metamerism

- **Two samples can look identical in one light, and the L*, a*, b* or X, Y, Z values can be identical, but can be completely different colors in another light. This is called metamerism.**

- **Metamerism can also be caused by surface difference interpreted as lightness difference.**

Ralph Stanziola

During the firing processes that chemically adheres the porcelain enamel to the steel substrate the glass again becomes molten and additional physical and chemical changes occur.

Changes that are occurring during fire process

Physical Changes -	Chemical Changes -
HEATING	can include
Enamel bisque shrinks -	
Steel expands -	Reduction -
Enamel becomes liquid-	Oxidation -
Enamel viscosity changes -	Solution -
Cooling	
Enamel becomes solid -	Crystallization -
Enamel and steel contract-	Gas evolution -
at different rates!	

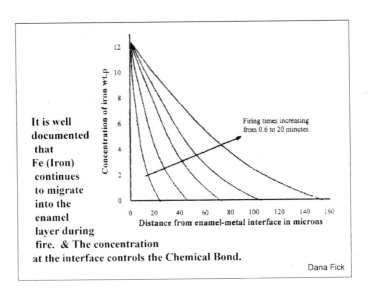

It is well documented that Fe (Iron) continues to migrate into the enamel layer during fire. & The concentration at the interface controls the Chemical Bond.

Firing times increasing from 0.6 to 20 minutes.

Concentration of iron wt.p

Distance from enamel-metal interface in microns

Dana Fick

Iron oxide (FeO) is dissolved in the glass during fire on steel; this iron oxide migration leads to another round of potentially color changing chemical reactions.

During fire of Porcelain Enamel:

- The steel oxidizes
- For chemical bond the oxide must dissolve into the glass
- There must be a high concentration of iron at the interface to generate good bond.
- If the fire is too low to allow for the oxide to dissolve or too high so as to change the equilibrium – then bond is weak.

Dana Fick

PE Powder Colors

- Improved color plans, earlier starts for product development –high cost
- Tighter control of processes at both the supplier and end user – high cost
- Inverse relationship volume - cost

While a wide range of colors are possible for PE Powder the economics of scale and degree of difficulty in maintain tight color controls limits the true demand for a range of colors.

POWDER ENAMELING CONTROL SYSTEMS

Phil Flasher
ITW Gema

By now everyone in manufacturing has been exposed to terms of lean manufacturing such as six sigma continuous process improvement, just in time manufacturing, pull manufacturing methods and zero inventory systems. For some, these terms are synonymous with their day to day jobs. If companies are not using this line of thinking then they are likely on the path to losing business to companies that are, or companies operating offshore in labor markets.

Improving the efficiency of a powder coating line is very important. And those companies taking the time to investigate and implement new tools and automation concepts will find their products being produced with a superior looking finish while realizing improved performance in the finishing line operational costs. Many users of powder equipment have taken the time to evaluate new spray gun technology, recovery booth technology and automation and have purchased those items in which they realized many benefits like improved film control, reduced material usage, lower operating costs and extended equipment life. This paper reviews some of the tools available and necessary to automate and control the application process.

Process control begins with the operator interface. Ease of use and intuitiveness is a must otherwise the given tools are frequently not used. Each gun control is interfaced with a PLC or PC based control system and has the ability to store all spray parameters for 255 recipes and more. Precise volumetric control of the air circuit to the injector creates repeatable powder deliveries from recipe to recipe. Electrostatic controls are also part of the recipe and are a must for enamel powder process control. High voltage and current controls can and will vary widely from coating mostly flat objects to the most complex oven cavities. All this is possible with gun controls like the Opti-star gun control. When implemented correctly these tools result in reduction of rejects, applied materials and manual operator intervention.

Other controls involve automatic movement and placement of guns to reduce or even eliminate the need for manual spray operations. These controls can involve reciprocation and up to six different axis of movement at one station in the most automated spray systems. The combination of different axis guarantees the correct positioning of each individual gun relative to the object being coated. The axis's feature toothed belts for extremely quiet operation with high stability even with heavy loading.

The porcelain enameling process for each object in its entirety is stored within the operator interface controls and can be connected to external process control and data collection systems for further quality control implementation.

Some companies take the wait and see approach while others are actively involved in a continuous improvement program. In today's global manufacturing market the company that takes a proactive approach to lean manufacturing will be around to see what happens in the end.

Once again, improving the efficiency of a powder coating line is very important. And those companies taking the time to investigate and implement any of the new tools and automation concepts will find their products being produced with a superior looking finish while realizing improved performance of the finishing line. But those that choose to do nothing in regards to automation will continue to find their operational cost spiraling out of control.

CUSTOMER EXPECTATIONS

Flexibility:
- Mixed production of different styles of parts to be enameled as well as different types of enamel
  - Combination of ground coat/topcoat and ground coat only

Cost reduction:
- Precise coating of semi-flat surfaces and deep drawn sinks or even oven cavities.
  - Increase of first pass yield to near 95%
  - Minimize film thickness variations and related rejects

Minimum reject rate:
- Process repeatability
  - Process stability
  - Automatic program change in between different products
  - No manual intervention
  - User friendly programming interface (Graphic Touch Panel)

Quality surface finish:
- No orange peel
  - Even distribution of speckled pyrolytic powders

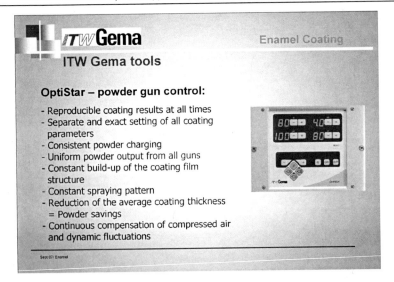

ITW Gema — Enamel Coating

ITW Gema tools

OptiStar – powder gun control:
- Reproducible coating results at all times
- Separate and exact setting of all coating parameters
- Consistent powder charging
- Uniform powder output from all guns
- Constant build-up of the coating film structure
- Constant spraying pattern
- Reduction of the average coating thickness = Powder savings
- Continuous compensation of compressed air and dynamic fluctuations

Sept 07/ Enamel

ITW Gema
Enamel Coating

ITW Gema tools

OptiGun – powder gun with main features:

- Special design for enamel powder
- High charging efficiency by using flat spray nozzle with central electrode
- wide range of nozzles for specific applications as:
  - deflector
  - deflector plate
  - flat spray
  - angle nozzle

Sept 07/ Enamel

ADVANCED PORCELAIN ENAMEL COATINGS WITH NOVEL PROPERTIES

Charles Baldwin and Jim Gavlenski
Ferro Corporation

ABSTRACT

Three new coatings have been developed which take today's stylish kitchen and bath finishes to a higher level, with performance that maintains beauty to build long-term brand popularity.

DURABLE METALLIC EVOLUTION™ FINISHES

Metallic surfaces, special effect colors, and reflective finishes have increasingly found their way into product design. The metallic trend shows no sign of slowing, but the growing demand for stainless steel and copper appliances, outdoor grills, fire bowls, and decorative items has dramatically boosted base metal prices.

For example, 304 stainless steel typically used for appliances contains 8 to 10.5% nickel. According to *USGS Minerals Information,* about two-thirds of all nickel production was earmarked for stainless steel in 2005. Cost data from the London Metals Exchange shows a steep price jump with nickel prices nearly quadrupling from $5.40/lb in December 2005 to $22.78/lb in May of this year. Copper prices have risen sharply as well.

Designers of appliances have taken these sharp cost increases on the chin, but that's about to change because of the economical Evolution™ line of porcelain enamels. These engineered coatings give trendy metallic looks with a substantially lower sticker price.

An estimate of cost savings per square foot for a typical kitchen range serves as an example. With Evolution™, designers can fabricate the range from 24-gauge enameling steel instead of 300 Series stainless. The coating would go on as a 0.002 in. (2 mil) ground coat and a 0.004 in. (4 mil) cover coat. Using general market prices from May '07, the metallic-look porcelain enamel could cut costs by nearly 60%. The porcelain coating also withstands scratches that would damage a significant percentage of stainless steel parts during fabrication, turning them into scrap.

Figure 1. Copper Evolution™ on a grill and Stainless-Style Evolution™ on a range

Evolution™ colors developed to date include several shades of stainless-look enamel, mirror finishes, and copper. Figure 1 shows examples of copper and stainless-style Evolution™. Oven interior pyrolytic ground coats with the copper color have also been created.

Evolution™ is applied and fired like conventional porcelain enamel. It provides the benefits of enamels such as:

- ✓ Sanitary qualities
- ✓ Easy cleaning
- ✓ Scratch and abrasion resistance
- ✓ Chemical and corrosion resistance
- ✓ Flame proof
- ✓ Color stability
- ✓ Environmentally friendly with zero solvents

Additionally, prior studies have shown porcelain enamels to be significantly more durable than 304 austenitic stainless steel.[2] Since austenitic stainless is more corrosion resistant than the ferritic stainless, one can conclude that enamels would be even more durable than the ferritics. Evolution™ colors are suitable for all parts of the appliance, housewares, and architectural markets. Evolution™ could also be used as accent pieces with stainless or for durable cabinetry.

STEAM-CLEANING AQUAREALEASE™

Oven cleaning technology consists of three types: (1) self-cleaning pyrolytic ground coat, (2) non-self-cleaning ground coat, and (3) catalytic continuous clean enamels. The first reduces foodstuffs to ash with exposure to temperature between 900 and 1000°F (482 and 538°C), while the second requires harsh alkaline cleaners to remove soils. The third has largely fallen out of use and relies on high-metals, porous enamels to catalyze the reduction of soils to ash at normal cooking temperatures.

AquaRealEase[TM] is a new alternative. It is porcelain enamel with a patented formulation that allows baked-on food residues to be released with exposure to moisture (either as water or steam). As such, it offers a more environmentally-friendly option to cleaning ovens without the use of harsh chemicals, high temperatures, and the resulting fumes. Additionally, interlocks and extra insulation would not be required, freeing the oven manufacturer to add other features such as advanced electronics or integration into a home network to the previously old-fashioned range.

AquaRealEase[TM] can be applied in a single coat over existing oven ground coats. It has the mechanical durability and thermal resistance of traditional enamel, can be applied in a single fire to steel, and fires out between 1470 and 1570°F (799 and 854°C). Figure 2 shows a photograph of a commercially produced range with an AquaRealEase[TM]-coated cavity.

Figure 2. Commercially-produced AquaRealEase range[3]

Although cleaning cycles may vary for various range manufacturers, AquaRealEase[TM] sheds baked-on soils in service with exposure to steam. The steam can be generated with closed vents by using only the lower heating element to heat 1 L of water in a pan on the lower rack to 190°F (88°C) for at least 30 minutes. Then, the oven is turned off and allowed to cool for 30 minutes. The sides of the cavity are wiped with the soft side of the sponge to allow water to run down the sides. After 20 minutes, the soils can be wiped out of the range. Because of the use of steam for the cleaning cycle, AquaRealEase[TM] has promise for steam cooking ranges emerging on the market.

NON-STICK REALEASE[TM] ENAMEL

RealEase[TM] is a state-of-the-art ceramic composite combining the cleanability of the organic non-sticks with the durability of vitreous enamel. It offers the hardness of enamel with the cleanability of PTFE.

Some of the benefits of RealEase[TM] include[4]:

- ✓ Cleanability equal to PTFE coatings and superior to stainless steel
- ✓ A pencil hardness of 8H
- ✓ Superior heat resistance compared to PTFE coatings

✓ FDA-conformable
✓ Environmental friendliness with zero solvents

While there continue to be significant concerns about the chemical components of PTFE-based non-sticks,[5] RealEase™ offers a vitreous ceramic finish that meets EPA 2010 guidelines.

RealEase™ can be applied to a wide variety of substrates:

✓ Aluminum
✓ Stainless steel
✓ Aluminized steel
✓ Cast iron
✓ Enameling steel
✓ Ceramic

For aluminum and stainless steel, roughening with blasting is required, while aluminized steel only needs an alkaline degrease. For cast iron and steel, the light application of a RealEase™ primer is necessary. Additionally, facilities with enameling capabilities, several vitreous primers called hard-bases are available, which have been used as base coats on ceramics.

Considering its properties, potential applications for a wipe-clean porcelain enamel usable up to 500 to 600°F (260 to 316°C) can be considered. For domestic uses, these include cookware and bakeware (aluminum, aluminized steel, stainless steel, or ceramic), small appliances, toaster and microwave ovens, simmer plates, and outdoor/backyard grills/griddles. Because RealEase™ is certified as safe for use in restaurant kitchens, it is also suitable for commercial kitchenware, cookware, bakeware, and appliances.

Figure 3. RealEase™-coated toaster oven

Figure 3 shows a commercially-available toaster oven with RealEase™ on the oven cavity. RealEase™ could also offer the same tough non-stick properties on free-standing or built-in ranges.

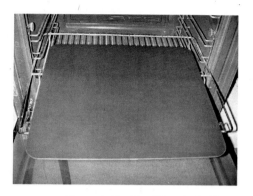 

Figure 4. RealEase™-coated baking tray and small appliance turntable

Additional current RealEase™ applications are shown in Figure 4 which are a release coating on a baking tray for the oven interior and on a turntable for a high-speed combination light-cooking/microwave oven.

SUMMARY

Three new enamel technologies were reviewed that offer new properties and applications: Evolution™ metallic-colored enamel cover coat, water-cleaning enamel AquaRealEase™ and non-stick enamel RealEase™. First, Evolution™ provides trendy finishes that are very suitable for high-end products. Second, AquaRealEase™ offers an oven-cleaning mechanism free from the necessity of high heat and fumes. Third, with a firing temperature and cleanability similar to PTFE coatings but superior mechanical and thermal resistance, RealEase™ would be most suitable for use on pots, pans, bakeware, griddles, and small appliances. Because of the scratch resistance of the material and a lower tendency to discolor on exposure to heat, the original appearance would be maintained longer than PTFE.

REFERENCES

[1] C. Baldwin and Tom Poplar, "Metallic Looks that Last," *Machine Design*, In Press.

[2] Dave Fedak and Charles Baldwin, "A Comparison of Enameled and Stainless Steel Surfaces," *Proceedings of the 67th Porcelain Enamel Institute Technical Forum*, 45 – 53 (2005).

[3] "Porter and Charles" http://www.porterandcharles.ca/ (6 September 2007).

C. Baldwin et al., "Advanced Coatings for the Home of Tomorrow," Appliance, Dec. 2006, pp. 20-24.

[5] "Chemicals in non-stick pans may retard babies' growth" http://news.independent.co.uk/health/article2896195.ece (6 September 2007).

ABSORPTION AND EMISSIVITY OF RADIANT ENERGY BY PORCELAIN ENAMELS

William D. Faust
Ferro Corporation

ABSTRACT

Porcelain enamel coatings can be tailored to provided specific absorptivity and emissivity properties to achieve improved thermal performance in a variety of applications such as solar heating and cooking applications.

INTRODUCTION

Heat transfer may occur by conduction, convection or radiation. We are familiar with each type in our daily lives. We can feel conduction when holding a warm cup of coffee, convection when we are warmed by air flowing past us and radiation when we are out in the sunshine. Heat transfer by radiation occurs naturally with all objects. There are two important considerations regarding radiation heat transfer, emissivity and absorptivity.

Emissivity is the rate at which radiation is given off from an objects surface. The radiation is essentially *infrared* radiation, that is, radiation between 1 μ-micron and 20 μ-microns[1.5]. Absorptivity is the absorption of infrared radiation by an object and is the reciprocal of an objects emissivity.[1] Absorption values for porcelain enamels were reported in the late 1970's in association with work on solar absorbing surfaces. Porcelain enamels were reported to have performances nearly identical to black chrome surfaces[2], one of the better absorbing surfaces being tested at that time.

The absorption and emissivity of porcelain enamel coatings has been primarily associated with solar energy applications and heat transfer by radiation.[2,3] Values for absorption and emissivity are comparisons to an ideal radiator, a black body. A black body absorbs all radiation incident upon it and reflects or transmits none. As an emitter, a black body radiates the maximum possible thermal radiation at all possible wavelengths.[1] At thermal equilibrium; the absorptivity and emissivity are equal. [1]

Conductors such as metal and non-conductors such as wood, paper, oxide films (glass) generally exhibit different emissivity properties. Metals, particularly un-oxidized surfaces, exhibit low emissivity values. Non-metals generally exhibit high emissivity values. The emissivity varies as a function of temperature and the associated wavelength of the thermal radiation.

Porcelain enamel coatings exhibit absorption and emissivity characteristics associated with glasses. Modification of the glass color and surface can affect these properties, but the effect on emissivity is less pronounced than reflectivity.

The source of radiation or heat energy must also be considered. Solar radiation is primarily concentrated between 0 and three milli-microns in wavelength with most of the radiation between 0.4 and 2.0 milli-microns. The solar radiation has the effect of temperature of 10,000°F (~5,537°C). As the temperature is reduced relative to solar radiation, the radiation occurs at longer wavelengths. At 212°F (100°C) or less, 90% of the radiation energy is between 3 and 30 microns. Figure 1 illustrates an ideal black body regarding radiation. With self-cleaning ovens, the heat source is typically about 1500-1600°F (815-871°C) which is estimated to be about 3 milli-microns.

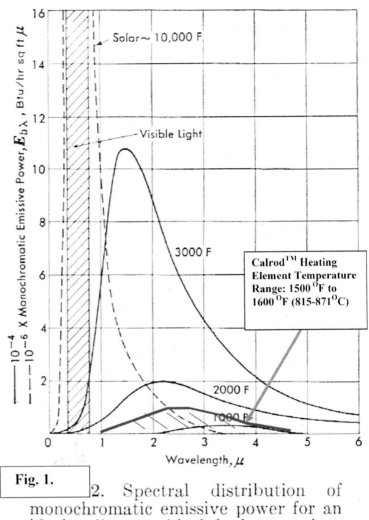

Fig. 1.

2. Spectral distribution of monochromatic emissive power for an ideal radiator or black body at various temperatures.

Figure 1: Black Body Radiation[1]

ABSORPTION

Absorption of energy by a surface is affected by its reflectivity. Light coatings which reflect a wider range of the solar spectrum heat at a slower rate than coatings that absorb most of the incoming radiation. The effect of color of the enamel on heating was reported by Baldwin[4] in 2000 (Figure 2) as a feature of more rapid heating of food due to the use of high reflectivity enamel in a heated enclosure.

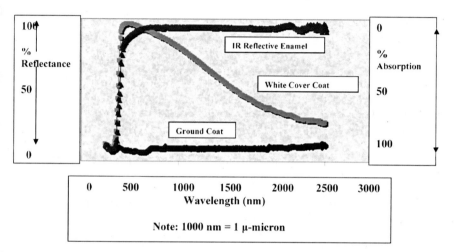

Figure 2: Reflectance Curves

Reflection values, the reciprocal of absorptivity, reported by Ruderer in 1977 show similar effects of high absorption of enamels on aluminum and steel as seen in Figures 3 and 4. The absorption values were obtained by integrating the reflectivity (reciprocal of absorption) values over 32 wavelengths.[3] The use of a very thin milli-micron thickness overcoat of a pyrolytically applied tin oxide doped indium oxide coating changed the thermal radiation characteristics of the coating drastically as seen in Figure 5. The surface layer of most solid objects, less than 0.05 inches (~1 milli-micron in thickness), controls the infrared thermal characteristics.[1]

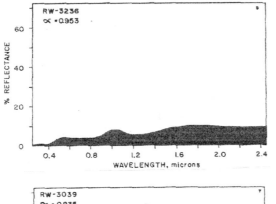

Figure 3

High Absorption (Low Reflectivity) Enamel Coating. Olive Drab Color, Aluminum Metal Substrate

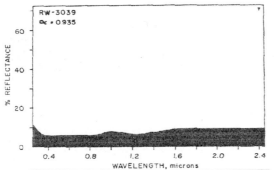

Figure 4

High Absorption (Low Reflectivity) Enamel on Aluminum Plate

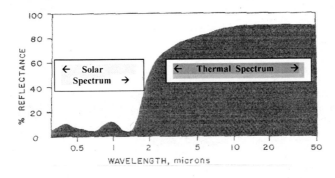

Figure 5 -Solar Selective Coating of Indium Oxide Doped with Tin Oxide on A Ground Coat Showing the Solar and Thermal Spectrums

EMISSIVITY

Emissivity measurements were carried out with a Devices & Services Company Emissometer Model AE. This instruments measures emissivity values at 180°F (82°C) using a high emissivity and a low emissivity standard.

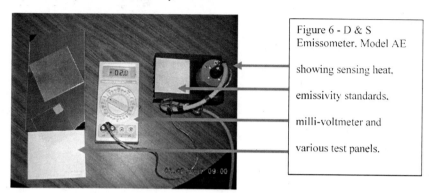

Figure 6 - D & S Emissometer. Model AE

showing sensing heat,

emissivity standards,

milli-voltmeter and

various test panels.

Emissivity Values for Various Surfaces Measured with D & S AE Emissometer

Glass, Clear Window Glass Section	0.85
Aluminum Metal, Shiny	0.04
"Silver Paint" on Top of Porcelain	0.39
"Gold Paint" on Top of Steel	0.54
Steel, Clean Matte Metal	0.04
Steel, Rusted	0.85

Emissivity Values for Porcelain Enamels Measured with D & S AE Emissometer

Ground Coat, Blue-black	0.85
Cover Coat, Titanium White	0.85
Pyrolytic Enamel	0.85
"Metallic Copper" Cover Coat	0.85
"Metallic" Appearance Enamel	0.85

Emissivity Values for Organic Powder Paints Measured with D & S AE Emissometer

White Coating	0.85
"Copper" Appearance Coating	0.85
"Silver" Appearance Coating	0.78
"Gold" Crinkle Coating	0.87

Emissivity Value for Solar Selective Porcelain Enamel Measured with D & S AE Emissometer

Ground Coat With Indium Tin Oxide Doped Interference Film Coating	0.33

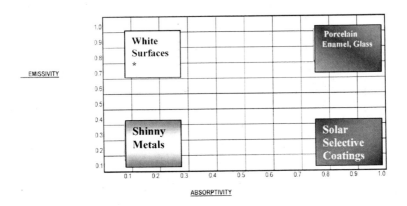

Figure 7 - Diagram of Absorptivity and Emissivity, * Porcelain Enamels, Glazes, Paints and Organic Powder Surfaces

DISCUSSION

Porcelain enamel coatings can be tailored to have specific absorptivity and emissivity characteristics as illustrated in Figure 7. High reflectivity coatings such as whites and dark low reflectivity coatings such as blue-blacks or blacks can be easily manufactured. A interference film coating layer has been shown to be industrially feasible.[3] Improved heat gain has been experimentally demonstrated on solar heated housing and in experimental cooking ovens.[3,4] Both of these applications have long term potential regarding durability and performance over many years of use.

REFERENCES

[1] Kreith, Frank, *Principles of Heat Transfer*, 2nd Edition, International Textbook Company, 1967, Chapter 5 - Heat Transfer by Radiation, page 198-215.

[2] Eppler, Richard A., "Solar Heating and Cooling Equipment," Proceedings of the Porcelain Enamel Institute Technical Forum, Vol. 38, 1976, pages 118 – 124.

[3] Ruderer, Clifford G., "Solar Update – Observations on Technical Advances," Proceedings of the Porcelain Enamel Institute Technical Forum, Vol. 39, 1977, pages 25-33.

[4] Baldwin, Charles A., "Heat Reflective Enamel for New Ovens," Proceedings of the Porcelain Enamel Institute Technical Forum, Vol. 62, 2000, pages 71-80.

[5] The Physics of Foil, Heat Gain/Loss in Buildings, Innovative Insulation, Inc., www.radiationbarriers.com/physics_of_foil.htm

NEW PRODUCT DESIGN, DEVELOPMENT, VALIDATION, AND LAUNCH: GETTING IT RIGHT THE FIRST TIME!

Kara Joyce Kopplin
QTEC Consulting

Successful product launches don't happen by accident. In order to launch the right product at the right time, many companies utilize a formal launch system consisting of specific task checklists, defined launch phases, and formal gate reviews between the phases. By taking a disciplined approach, companies can minimize costly blunders.

It's a long road from idea generation to a full production launch, and that road is fraught with risks – to the consumer, to the manufacturing employees, and to the manufacturing company. Product Development Processes are effective in minimizing those risks. PDPs are structured in phases (typically five), and on completion of each phase, a formal committee reviews all the tasks and deliverables. These reviews ensure customer needs are met, regulatory requirements are met, and the project continues to be a good fit for the company. Five phase PDPs are typically structured as follows:

PHASE 1. PROJECT JUSTIFICATION – BUSINESS CASE

An idea is submitted to the development committee. Their review is the first gate. If the idea has merit, it enters phase I. In this phase, the idea is analyzed by researching user needs and expectations, the potential market, potential risks, required resources, technical and administrative requirements, and compatibility with the company's strategic goals. The second gate is a review of the business case, and a successful proposal leads to the next phase:

PHASE 2. PRODUCT DEFINITION – CUSTOMER NEEDS

Specific requirements are determined in this phase, by working with the customer, sales, and the engineering group. Product details are specified, including function, appearance, reliability, and maintenance. The location of product use is defined, such as environmental conditions, compatibility with chemicals or other products, and the human interface. Standards and regulations specific to the product or the market are defined, and the target manufacturing costs determined.

These first two phases are inputs to the design process, and as such, determine the resulting outputs. The challenge is to take time to gather all the pertinent information and ensure all details are understood. Gate three is particularly important to prevent leaping into Phase 3 before all the required inputs have been agreed upon.

PHASE 3. DESIGN AND DEVELOPMENT OF PROTOTYPE

Based on the inputs, the design team develops a prototype. The gate between Phase 3 and 4 is often internal to the R&D group, as they develop prototypes, test and validate, then modify designs with new prototypes.

PHASE 4. TESTING AND VALIDATION

Once all verification and validation testing is complete, risk analysis has been performed, and the prototype outputs clearly meet the inputs, the prototype is reviewed and approved by the committee in gate five.

PHASE 5. LAUNCH

Phase 5 is often very extensive since operations launches require numerous tasks to be completed by several departments. Typically companies need to:

Manage any remaining risk identified during risk analysis, develop the process, qualify equipment, set process parameters and validate processes, assign part numbers, set up inventory, make purchasing arrangements, establish quality requirements, develop testing, define record requirements, create work instructions, labels, and operations manuals, develop packaging, increase staffing, perform training, develop final sales figures and distribution channels, and launch the marketing and advertising campaigns. Clearly a task list is helpful in tracking all of these key activities!

Once all operations activities have been completed, a final gate review is held to begin production.

Gate reviews between phases ensure executive management remains informed about project progress. Project funding can be released in steps, and the direction of a project can be halted or re-directed before the focus creeps. If such reviews are not held periodically, the focus of the project can shift far from what the customer wanted, after the company has made investments in development time, tooling, and materials. Checking the details early and often prevents costly missteps.

After a product has launched, typically six months or a year later, a post-mortem can be valuable to create a history that can be used to guide future projects or improve the current production. It helps the team figure out what went well and what didn't, so they can learn from the successes and shortcomings. The knowledge can be used to improve the Product Development Process.

Besides making good business sense, the PDP can satisfy the requirements of many quality management system standards, including ISO9001:2000, TS16949, and ISO13485. Tools that can be used within the process include design and process failure mode effects analyses (DFMEAs and FMEAs), risk management (as guided by ISO 14971), and design for six sigma.

Good processes yield good products. Using PDP provides a framework that takes a project step by step from the initial idea to the product launch through discipline, teamwork, and leadership. PDP provides a structure that guides the activities and interaction of all the departments. It utilizes checks and balances to keep product development from straying from the goal.

PDP prevents resource misallocation, and minimizes the risk that products will not meet the intended use, are unsafe or ineffective, do not meet customer requirements, or are not profitable.

The ultimate value of using a PDP is to launch safe, effective products on time and under budget, that will delight the customer.

UPDATE ON CURRENT EPA AND OSHA ISSUES THAT WILL IMPACT YOUR INDUSTRY

Jack Waggener
URS Corporation
Nashville, Tennessee

ABSTRACT

A final rule by OSHA (Occupational Safety And Health Agency- USA) has been issued regarding hexavalent chrome (Cr^{+6}) in dusts, mists and fumes which will impact many industrial processes that use green chrome oxide, welding systems, paints using chrome sealers, electroplating, and other processes. This rule is in many instances more stringent than those found in numerous other industrial countries.

Silica exposure limits are being reviewed with the possibility of reduction in the exposure limit to about 50% from the current level of 100 micrograms per cubic meter of air. Porcelain enameling facilities have been removed from the EPA listing for waste water nitrite effluents which readily convert to nitrates. Future restrictions on ozone generation and other greenhouse gases (GHG) such as methane, nitrous oxide and fluorinated gases are considered. The new European Union "REACH" policy [Registration, Evaluation & Authorization of Chemicals] will have a global impact as it is phased in over the next 10 years.

Permissible Exposure Limit (PEL) (8-hr TWA)

Past PEL	52 ug/m^3
Proposed PEL	1 ug/m^3
Action Level (AL)	0.5ug/m^3
Final PEL	5.0 ug/m^3
AL	2.5 ug/m^3

Low PEL
Impacts Many Processes in P.E. Plants & Suppliers

Cr Oxides (PE)
Welding SS
Cr Sealers (Paint Lines)
Cr Electroplating
Welding Mild Steel
Chromates
Polishing/Grinding
Anodizing

Cr^{+6} PEL
Occupational Exposure Limits:
Comparison of Selected Countries
(Trading Partners)

Country	Occupational Exposure Limit
United States	
OSHA Final	5.0 ug/m3
OSHA Past	52 ug/m3
Japan	50 ug/m3
European Union	50 ug/m3
France, Germany, UK, Finland	50 ug/m3
China	50 ug/m3
India	50 ug/m3
Sweden	20 ug/m3
Denmark	5 ug/m3

OSHA
Respirable Crystalline Silica
PEL

Current PEL   **100 ug/m$^{3(\pm)}$**

PEL's under consideration

PEL	**Action Level**
100 ug/m^3	50 ug/m^3
75 ug/m^3	40 ug/m^3
50 ug/m^3	None

NIOSH & ACGIH Recommends 50 ug/m^3

Waste Water

EPA Proposes (8/29/05) to evaluate revising the Porcelain Enamel Effluent Limitation Guideline (1982)

2006-PEI meets with EPA & Identifies EPA mistakes

2007-EPA removes PE from list

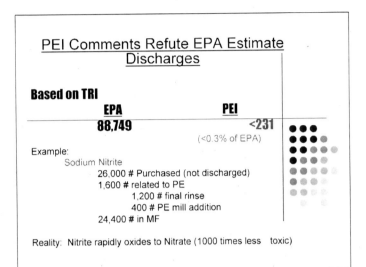

PEI Comments Refute EPA Estimate Discharges

Based on TRI

EPA	PEI
88,749	**<231**
	(<0.3% of EPA)

Example:
Sodium Nitrite
26,000 # Purchased (not discharged)
1,600 # related to PE
1,200 # final rinse
400 # PE mill addition
24,400 # in MF

Reality: Nitrite rapidly oxides to Nitrate (1000 times less   toxic)

14 PE Plants Evaluated

American Standard, Salem, OH	TRI
Electrolux, Springfield, TN	TRI
Hanson Porcelain, Lynchburg, VA	TRI
Kohler, Cast Iron, WI	TRI
Maytag, Newton, IA	TRI
Maytag #1 & #3, Cleveland, TN	TRI
Roper, Lafayette, GA	TRI
VITCO, Nappanee, IN	TRI
Whirlpool, Clyde, OH	TRI
Whirlpool, Tulsa, OK	TRI
State Inds., Ashland City, TN	TRI & PCS
Briggs, Knoxville, TN	PCS

AIR
OZONE

- July, 2007 EPA Proposes **REDUCTION**
  - Primary Standard (Human Health)
  - Secondary Standard (Vegetation/Crops)
- Summer, 2008 Rule Effective
- Big Impact on Most URBAN Areas
  - Painting
  - Solvent Use
  - Emissions

"REACH"
(Registration, Evaluation, & Authorization of Chemicals)

- New "EU" Chemical Policy

- June 1, 2007 (Initiation)
  Next 10 years: Phased Approval

- Impacts imports to EU

- Goal: "REACH be used globally"

CASE STUDY: BENEFITS OF A NEW HIGH EFFICIENCY FURNACE

Jason Butz
Engineered Storage Products Company
DeKalb, IL

INTRODUCTION

Engineered Storage Products Company makes porcelain-enameled steel bolt-together storage tanks. They make tanks for agricultural, municipal, and industrial applications. These tanks come in a variety of sizes, and use low carbon steel sheets that vary in thickness from 0.080 inches up to 0.500 inches.

In 2005 Engineered Storage Products Company realized that they needed to increase the throughput of their furnace in order to keep up with product demand. It was determined that they could either improve the efficiency of the old furnace by refurbishing it, or purchase a new high efficiency furnace. A project team was then assembled in order to evaluate their options. After comparing the cost of refurbishing the old furnace to that of replacing it with a new furnace, it was determined that it would be less time consuming and more cost effective to replace the existing furnace with a new high efficiency furnace. Not to mention that the old furnace was at the end of its life, and would not be as highly efficient as a new furnace even if it was refurbished.

OLD FURNACE

The old furnace started out as a Ferro electric porcelain enameling furnace, and was originally built in 1975. In January 1982 it was converted into a gas furnace by Can-Eng Ltd. because electricity cost had started to rise at a much higher rate than natural gas.

The furnace was a straight-through furnace which had a total of 33 burners throughout its three different zones. The hot zone of the furnace was 114 feet long and reached a temperature of 1600° F. After exiting the furnace, the product entered the cooling tunnel which consisted of a water-jacketed cooling tunnel followed by a forced air cooling tunnel. The line speed was 12 ft/min for 0.094 inch thick steel, which represents the highest percentage of sheets run. The furnace had a throughput of 27,000 pounds of steel per hour and had a 2-½ hour start up time.

NEW FURNACE

The new furnace is a high efficiency gas furnace supplied by KMI Systems, Inc. and it was installed in December of 2005 for a total project cost of approximately $1.8 million. It is currently the largest porcelain enameling furnace in North America.

The furnace is U-shaped in design, which allows it to be more evenly heated from top to bottom through out the hot zone. This design also allows for the incoming sheets to be heated and the outgoing sheets to be cooled as they pass by one another. The furnace has a total of 34 high efficiency burners throughout five different zones. The hot zone is 120 feet long and reaches a high temperature of 1600° F. After exiting the furnace, the product enters a forced air cooling tunnel. The line speed is 14.5 ft/min for 0.094 inch thick steel, which again represents the highest percentage of sheets run. The furnace has a throughput of 40,000 pounds of steel per hour and has a 1-½ hour start up time.

BENEFITS

Engineered Storage Products Company determined it would be more beneficial to install a new furnace rather than refurbishing the existing one, because they could get a much higher throughput with greater efficiency from a new furnace. There were also other benefits such as the reduction in operating costs and product quality improvement that contributed to their decision that a new furnace was the best option.

Throughput

The new high efficiency furnace, as stated before, has a throughput of 40,000 pounds of steel per hour. This is an increase of just over 30% in steel throughput when compared to the old furnace, which had a throughput of 27,000 pounds of steel per hour. This is in part due to the fact that when firing 0.094 inch thick steel sheets, the furnace line is able to run at 2-½ ft per minute faster than when running the same sheets through the old furnace.

Another reason for the increase in throughput is that the old furnace was limited by the amount of heavier thickness sheets that were able to be hung on the line. The thicker the steel, the more time at temperature that is needed to heat the sheet to properly fire the glass. Sheet thicknesses ranging from 0.080 inch - 0.187 inch were able to be hung solid on the furnace line. However, any sheets thicker than 0.187 inch steel were hung by hanging two sheets and skipping one hanger before hanging two more sheets. Currently, the sheets do not have to be hung, hang two skip one, until they get to 0.312 inch thick steel (See Fig. 1). Because of this, the line is now able to handle about 95% of the total sheets being fired with out having to skip any hangers. The old furnace line could only handle about 80% of the total amount of sheets being fired with out having to skip any hangers.

Sheet Gages	Old Furnace Hanging Pattern	New Furnace Hanging Pattern
0.080	Solid	Solid
0.094	Solid	Solid
0.125	Solid	Solid
0.156	Solid	Solid
0.187	Solid	Solid
0.218	Hang 2 Skip 1	Solid
0.250	Hang 1 Skip 1	Solid
0.281	Hang 1 Skip 1	Solid
0.312	Hang 1 Skip 1	Solid
0.344	Hang 1 Skip 1	Hang 2 Skip 1
0.375	Hang 1 Skip 1	Hang 2 Skip 1
0.438	N/A	Hang 1 Skip 1
0.500	N/A	Hang 1 Skip 1

Fig. 1: hanging pattern for old and new furnace

Efficiency

The efficiency of the furnace is due to a few things. First, its U-shape design helps to make it efficient, because as the hot parts come out of the furnace and pass by the cold parts entering the furnace heat transfer takes place between them. This causes the cold parts to heat up to temperature quicker, while simultaneously cooling the hot parts. By reducing the amount of time and temperature that is required to heat the sheet, the amount of gas that is being used is reduced, therefore making the furnace more efficient. With the old straight-through furnace the sheets never passed by each other so heat transfer was unable to take place, which means the burners must work harder to get the sheets up to temperature and more gas is used.

The burners are a standard S-tube design, and they empty into a collector duct running along the furnace floor to the preheat area. This duct collects the waste heat and sends it to the preheat area, where the heat is used to help bring the sheets up to temperature quicker.

The U-shaped furnace design is also more efficient because it helps to keep the temperature more regulated from top to bottom in the hot zones throughout the furnace. This can be shown by comparing steel temperature differences from a Datapaq run through both the old and new furnaces (See Fig. 2a, 2b). The temperature difference in the hot zones of the old furnace are 40 - 90°F top to bottom, whereas in the new furnace the temperature differences are 5 - 15°F top to bottom.

**Notice the Difference in
temperature from top to
bottom is greater with the
old furnace**

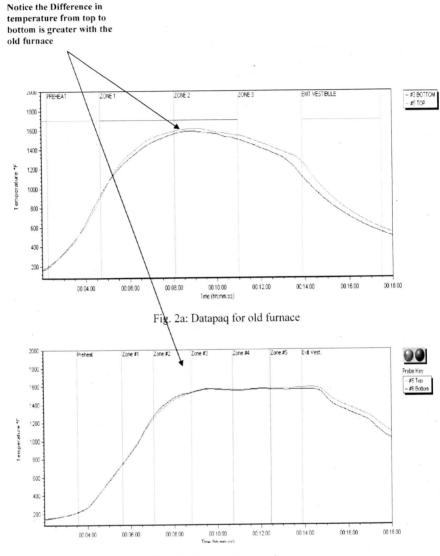

Fig. 2a: Datapaq for old furnace

Fig. 2 b: Datapaq for new furnace

Operating Cost Reduction

The operating costs were greatly reduced, not only through efficiency, which translated to lower gas usage to run the new furnace, but also in maintenance cost. The old furnace was in need of maintenance quite frequently, and this cost around $30,000 a year. For example, the S-tubes in the furnace had deteriorated to the point where they could no longer be welded up and had to be replaced on a frequent basis. Also, the water jackets in the cooling tunnel would often rupture, shooting water into the end of the furnace. This caused a problem not only because the water jacket needed to be repaired, but also because this would cause unwanted moisture to enter the furnace, causing defects in the products being run. Additionally, the pumps used for the water cooling frequently broke down and needed to be repaired.

The water-jacketed cooling tunnel also required the company to get testing done and to get a permit every year. These items cost the company $12,000 a year.

The money saved by the company in maintenance and permits alone was huge. Another area where the company was able to save money was quality improvement.

Quality Improvement

The new furnace drastically improved the quality of the product while saving the company money. The most troublesome problem that the new furnace fixed was holidays (pinholes in the coating). The reject rate for holidays dropped significantly after the new furnace was installed. This was a savings because when a holiday was found, the sheet had to be sent around the line a second time to be resprayed. Instead of the process being the standard 3 coat 1 fire process, it became a 4 coat 2 fire process, which meant higher glass coating and gas usage. Additionally, more labor was required for those particular sheets.

CONCLUSION

In conclusion, one can see that the decision that Engineered Storage Products Company's project team made to replace the old furnace with a new high efficiency furnace was the right one. The new high efficiency furnace provided a number of significant benefits including:
- increased throughput
- better heating efficiency which translated directly into lower gas costs
- lower maintenance and permitting costs
- significantly reduced reject rate

These benefits clearly justified the cost of the new furnace. A new furnace should be considered as a viable alternative when looking at a major repair of an existing furnace.

CHEMICAL BONDING OF CONCRETE AND STEEL REINFORCEMENT USING A VITREOUS ENAMEL COUPLING LAYER

Larry Lynch, Charles Weiss, Jr., Donna Day, Joe Tom, Philip Malone
USAE Research and Development Center
Vicksburg, MS

Cullen Hackler
Glass Technology Development Corp.
Alpharetta, GA

ABSTRACT

The bond between the cement paste and steel reinforcement in concrete structures is typically very weak. Enameling the steel reinforcement with a low-melting point vitreous enamel that contains calcium silicates and calcium aluminoferrites raises the bond strength by a factor of approximately two to four times the normal bond adhesion. An X-ray diffraction study and SEM imaging and a microanalytical examination of the bonding enamel layer indicates the calcium silicates and calcium aluminoferrites fused into the enamel react with the moisture in the fresh mortar and form a calcium silicate hydrate surface that adheres to the surrounding paste. The increased bond strength demonstrates that the enamel-cement coating acts as a coupling layer and attaches the hardening calcium silicate gel layer to the steel. During hydration, the composite cement-glass layer reacts with water that accumulates at the interface and appears to prevent the usual weakly-cemented layer from occurring in the adjacent paste. Additionally the bonding enamel can be applied to provide a durable chemically-resistant layer to protect the steel from oxidation.

1. INTRODUCTION

When steel reinforcement is embedded in fresh concrete and the concrete is consolidated a saturated calcium hydroxide solution accumulates at the surface of the steel. As the cement hydration reaction progresses the calcium silicate gel that formed shrinks and a layer of calcium hydroxide (portlandite) is deposited on the surface of the metal and a water-rich, weak paste layer forms adjacent to the calcium hydroxide layer. The two layers taken together are approximately 50 μm in thickness and very non-uniform [1-3]. This interfacial transition zone is considered the weak link in the paste and any dense inclusions in the paste (aggregate, fiber, or reinforcing steel) [4]. Studies conducted on the effects of coatings and changes in the composition of the paste fraction of the concrete indicate that only minor increases in the bond strength can be obtained by additions such as silica fume [5] and some coatings may even reduce the interfacial bond strength [6,7].

The difficulty in developing bonding coatings for steel reinforcement in concrete can be related to the complex nature of the cement paste and the lack of intermediate phases that will perform as coupling agents. Hydrating portland cement paste is largely a mixture of calcium hydroxide, calcium carbonate, and calcium alumino-sulfate crystals in a calcium silicate hydrate gel [8, 9]. Over chemical compounds are thought to form during the hardening of cement pastes [10]. Investigations undertaken with steel reinforcement in conventional concrete indicate that a layer of soft crystals of calcium hydroxide are the most common material noted at the steel surface [11]. A dense calcium silicate hydrate phase typically does not form at the surface of steel reinforcement unless the composition of the concrete is altered to prevent segregation of the concrete at the iron-paste interface [5, 12, 13]. Even when a hardened paste can be produced around steel, the phases at the interface are usually ferrous and ferric hydroxides that are not tightly bonded to the silicate gel phase in the paste. Without a bonding layer on reinforcing steel, the best bonding mechanism that can be postulated is the production of an electrical double layer at the contact of the paste and the steel. Calcium, aluminum and silicon coupled by electrical charges across the interface with hydroxide ions on the surface of the steel and iron atoms couple with unbalanced oxygen atoms in the paste [14]. Mlodecki [15] describes the bond between the iron and the surrounding calcium silicate gel as a form of hydrogen bonding involving the iron and the hydroxyl groups in the gel. The interface of the cement paste and the iron is weaker than any boundary in the gel or the metal.

The present investigation examines effects that can be produced by fusing modified porcelain enamels containing the hydraulically reactive phases in Portland cement to a steel surface prior to embedding the steel in concrete. This approach is different from previous treatments in that it can firmly bond a reactive silicate phase to the steel and develop an outer layer of enameling glass and cement grains capable of bonding to the surrounding hydrating paste. SEM, microanalytical, and X-ray diffraction techniques are employed to characterize the enamel coating before and after cementing with a standard Portland cement mortar.

2. MATERIALS AND METHODS

2.1. Preparation of Test Specimens

Test specimens consisted of 12.7-mm wide by 1-mm thick mild steel strips cut to be 152.4-mm in length. One end of the strip was drilled to allow it to be attached to the test apparatus. The length of the strips permitted them to be embedded in mortar to a depth of 101.6-mm and permitted the collection of sufficient sample of surface and interface materials that allowed characterization using SEM-based and X-ray diffraction techniques.

The test strip surfaces were prepared for ground coat enameling using either an alkaline cleaning process or a grit blasting process. While an acid etch/nickel deposition preparation process has historically been used, it is not necessary using today's ground coat compositions designed for application to "cleaned only metal." This cleaned only surface preparation generally is a series of water rinses under controlled temperatures.

2.2. Composition of Frit

The composition of the glass frit applied to the test rods varied with the manufacturer and the exact composition of most formulations is proprietary. In all cases the manufacturer was asked to furnish an alkali-resistant formulation that would be a suitable ground coat for a two-firing application. The composition for a typical glass frit prepared for this application is given in Table 1.

Table 1. Composition Range of a Typical Alkali-resistant
Groundcoat Enamel for Steel [16,17]

Constituent	Amount (%)	Range
Silicon dioxide SiO_2	42.02	40 – 45
Boron oxide B_2O_3	18.41	16 – 20
Sodium oxide Na_2O	15.05	15 – 18
Potassium oxide K_2O	2.71	2 – 4
Lithium oxide Li_2O	1.06	1 – 2
Calcium oxide CaO	4.47	3 – 5
Aluminum oxide Al_2O_3	4.38	3 – 5
Zirconium oxide ZrO_2	5.04	4 – 6
Copper oxide CuO	0.07	nil
Manganese dioxide MnO_2	1.39	1 – 2
Nickel oxide NiO	1.04	1 – 2
Cobalt Oxide Co_3O_4	0.93	0.5 – 1.5
Phosphorus Oxide P_2O_5	0.68	0.5 – 1
Fluorine F_2	2.75	2 – 3.5

2.3. Application of Frit-Cement Mixture

Bonding enamel can be applied in a single coat or double coat process [18]. In the single coat process used in this project the bonding enamel was applied by making a slurry of frit and clay with water containing cement plus the necessary surfactants to achieve the desired suspension and viscosity. The test rods were coated by dipping or flow coating the slurry onto the surface. Portland cement (Type I-II) and frit were fired in a one-step process; the portland cement was mixed in a 50% proportion by volume with the frit. The cement was added to the porcelain enamel slurry and applied to the steel surface.

2.4. Firing of the Frit

The porcelain enamel coating was fired onto steel at temperatures from 745 to 850 °C. Firing times are typically from 2 to 8 minutes depending on the mass of metal to be heated and the size of the furnace. The goal was to produce a groundcoat enamel 50 to 100 μm (2 to 4 mils) thick with the cement embedded in the ground coat. No attempt was made to obtain an even or smooth coating as would normally be the case for porcelain enamels for appliances, bathtubs, and other enamel items (Fig. 1).

Fig. 1 Coated metal strip embedded in mortar cylinder after pull-out testing. Note strain resulted in fracturing the test cylinder; but mortar remained attached to the bonding enamel coating

2.5. Preparation of Test Mortar

The enameled test strips were embedded in a mortar prepared using the guidelines presented in ASTM C 109, Standard Test Method for Compressive Strength of Hydraulic Cement Mortars Using 2-in. (or 50-mm) Cube Specimens [19]. The proportion of the standard mortar was one part cement (Type 1-II, Lone Star Industries, Cape Girardeau, MO) to 2.75 parts of standard graded sand meeting ASTM C 778, Standard Specification for Standard Sand. The water-to-cement ratio was maintained at 0.485. Test cylinders were prepared for each mortar batch and tested to determine the unconfined compressive strength at 7 days was within the limits recognized for this mixture design [20].

2.6. Testing of Embedded Strip Specimens

Each enameled test sheet or strip was inserted in a 50.8-mm in diameter, 101.6-mm long cylinder mold filled with fresh mortar. The strip was clamped at the top so that a 101.6-mm length of the coated portion of the rod was in the mortar. Each cylinder was tapped and vibrated to remove any entrapped air and to consolidate the mortar. The samples were placed in a 100% humidity cabinet at 25 °C and cured for 7 days. After 7 days, the test cylinders were de-molded and the mounted in the test apparatus and the force required to pull the steel strip out of the mortar was measured using an MTS Model 810 Testing Machine (Material Testing Systems, Minneapolis, MN). The testing procedure followed the standard protocol outlined in ASTM A 944, Standard Test Method for Comparing Bond Strength of Steel Reinforcing Bars to Concrete Using Beam-End specimens [21]. Following testing, the strip samples were recovered from the test cylinders. The surface morphology and composition of the test samples was characterized as to morphology using a Philips ESEM Model 2020 with a cerium hexaboride (CeB6) electron source and a gaseous secondary electron detector (GSED). The imaging system used an accelerating voltage of 20 KeV and 1.81 mA, and approximately 5 Torr (666 Pa) water vapor in the sample chamber. Images of these samples were collected over a period of 30 seconds and stored as 1 MB TIF files. Microanalysis was performed using a Buker electron-excited X-ray fluorescence system attached to the ESEM.

Mineralogical analysis of samples was conducted using X-ray diffraction (XRD) analysis on random powder samples using standard techniques for phase identification. The equipment used in these analyses is a Philips PW1800 Automated Powder Diffractometer system. The run conditions include the use of CuKα radiation and step scanning from 2 to 65° 2θ with 0.05° 2θ steps, and collecting for 3 to 4 sec per step. The collection of the diffraction patterns is accomplished using Datascan (Materials Data, Inc.) and analysis using Jade.

3. RESULTS AND DISCUSSION

3.1. Pull-out Testing

The results of the pullout tests of the metal strip samples are presented in Table 2.

Table 2. Results of Pull-out Tests on Strip Samples

Sample No.	Peak Force (N)	Bond Strength (MPa)	Remarks
1	8301.4	3.22	Cylinder split
2	11180.7	4.33	Cylinder split
3	11023.5	4.27	Cylinder split
4	9461.3	3.67	Cylinder split
5	10677.6	4.14	Cylinder split
6	10401.7	4.03	Cylinder split, 10-mm extension (?)
Average	10174.5	3.94	
Std. Deviation	1100.3	0.43	

Published values for bare (uncoated) steel and coated steel in portland cement mortar are compared to the results of this study in Table 3. Note that the pull-out tests for the steel strips show the bond strength above the range typical for uncoated steel, but below the maximum obtained for bonding enamel-coated rods. The data from the coated and uncoated rods were obtained using 6.35-mm diameter rods embedded in 50.8-mm diameter mortar cylinders. The tests were completed without fracturing the test cylinders. The lower value observed for the steel strips may reflect the effect of the mortar strength and cylinder size than the bond strength.

Table 3. Comparison of Average Bond Strengths

Cement-Steel System	Average Peak Force (N)	Std. Deviation (N)	Effective Bonding Area (mm^2)	Average Bond Strength (MPa)
Steel fiber in mortar [19]	---	---	---	2.04 - 2.72
Smooth Rods, uncoated [23]	2618.2	466.2	1266	2.06
Rods with Bonding Enamel [23]	11124.6	235.3	1266	8.79
Strips with Bonding Enamel (this study)	10174.5	1100.3	2580	3.94

The increase in bond strength demonstrates that the reactive enamel hydrated and produced the same bonding effect previously observed with the rod samples.

Comparison of the pullout test data with published test results from coated and uncoated metal fibers and rods (Table 3) indicates the bond strengths are approximately two times greater than uncoated metal samples and within the range reported for metal rods coated with bonding enamel. All of the test cylinders released the metal strips by splitting down the length and the mortar was typically still adhering to the metal. The mode of failure suggests that the measured strength was related to the strength of the mortar mixture since most of the failures occurred in the mortar adjacent to the bonding layer and not at the metal surface. This suggests that higher pull-out forces might have been required if a higher strength mortar mixture has been employed in the test or larger diameter mortar test specimens.

3.2. Examination of the Morphology of the Coated Steel Surfaces

SEM images of metal samples cut from strips before hydration of the coating, after hydration of the coating, and after the coated strip in the mortar had been cracked during testing and the cemented section removed are shown in Figs. 2 through 5.

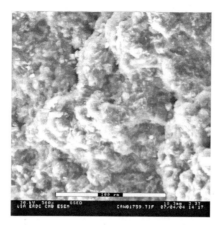

Fig. 2 SEM image of the surface of the unhydrated bonding enamel surface. The cement grains are embedded in the enameling glass, but the exposed surfaces can hydrate

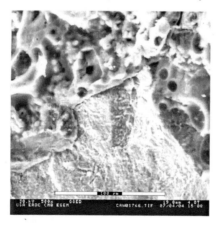

Fig. 3 Hydrated bonding enamel chipped to show the underlying metal with the remaining enamel glass

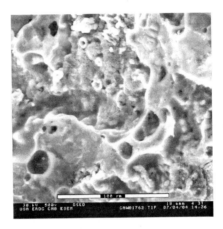

Fig. 4 Surface of hydrated bonding enamel. Note the surface crystal growth of calcite from the carbonation of portlandite produced by hydration

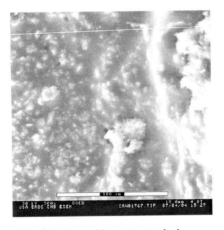

Fig. 5 Surface of cemented bonding layer with mortar attached

The surface of the unhydrated bonding enamel (Fig. 2) shows the portland cement grains embedded in the enameling glass. Portland cement clinker is fired at a temperature of 1480 to 1650 °C and the clinker grains are not altered during the enameling process. The sharp ends of the cement grains are evident in the photomicrographs. Fig. 3 shows a chipped portion of a hydrated glass bonding layer with the surface of the metal (with the remaining enameling glass on the surface). There is no evidence of a intermediate layer between the glass and the metal. The surface of the hydrated enamel is shown in Fig. 4. The crystalline material on the surface appears to be calcite form the carbonation of portlandite formed during the hydration of the cement at the enamel surface. The cemented enamel (Fig. 5) shows the same morphology as the hydrated enamel. Patches of mortar are bonded to the surface. There is no evidence of a separate layer between the mortar and the enamel-cement layer.

3.3. Comparison of the Microanalyses of Bonding Enamel

Table 4 summarizes the chemical composition data obtained from the X-ray fluorescence analyses of the unhydrated, hydrated, and cemented surfaces. Hydration involves adding water to create hydrated silica phases from the enamel-cement mixture and the compositions of the hydrated enamel samples shows this with an increase in the oxygen content. The cemented enamel is bonded to the mortar that consists primarily of calcium silicate hydrate gel, and quartz. The addition of the mortar is reflected in the higher silica and calcium content of the cemented enamel and the lower concentration of sodium in the cemented enamel sample compared to the hydrated sample. The decrease in sodium concentration from unhydrated samples to cemented samples reflects the decrease in the proportion of enamel glass in the sample.

Table 4. Chemical Composition of Bonding Enamel Samples

Element	Unhydrated enamel (mass %)	Hydrated enamel (mass %)	Cemented enamel (mass %)
O	45.56	51.33	51.16
Ca	28.28	29.51	31.40
Si	12.12	8.54	9.79
Na	3.06	1.88	0.67
Fe	4.45	2.87	1.91
Al	2.05	1.46	1.08

3.4. Comparison of the X-ray Diffraction Data

The X-ray diffraction patterns produced from random packed powder samples of each of the bonding enamel sample collected are shown in Figures 6 through 8. The unhydrated bonding enamel sample shows the peaks for the crystalline phases in the portland cement along with traces of calcite. Peaks for both anhydrite and gypsum are present. The firing temperature for the enameling is sufficiently high to dehydrate gypsum (over 360°C); but the time may be too short for the reaction to go to completion.

The hydrated bonding enamel (Fig. 7) shows a large peak for calcite, but there is no peak for portlandite (the most common crystalline hydration product). Portlandite is easily converted to calcite from contact with the carbon dioxide in the surrounding air. The cemented bonding enamel sample (Fig. 8) contains the bonding enamel glass-cement mixture and mortar from the test cylinder that adhered to the enamel. The quartz sand from the mortar produces a prominent peak. Peaks are present for ettringite, portlandite, and calcite. The cemented sample contains all of the crystalline phases that would be produced from hardened mortar.

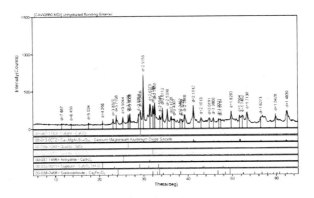

Fig. 6 X-ray diffraction pattern from unhydrated bonding enamel sample. Both gypsum and anhydrite are present

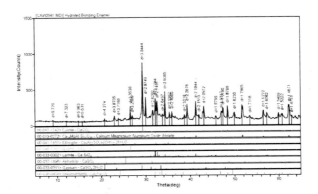

Fig. 7 X-ray diffraction pattern from hydrated bonding enamel sample. Calcite forms from carbonation of portlandite produced in hydration. Peaks consistent with normal cement hydration are present

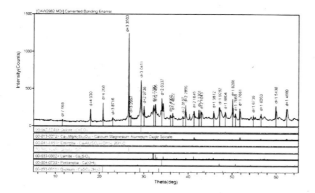

Fig. 8 X-ray diffraction pattern from cemented bonding enamel sample. The quartz peak is attributed to the mortar adhering to the cement-glass layer

4. SUMMARY

The investigation of the bond strength, morphology, micro-chemistry, and characteristics of the crystalline phases in the bonding enamel showed the following:

1. The cement-glass layer fused to the surface of steel produces a bond strength that exceeds that for comparable untreated steel surfaces. Failures during pull-out testing usually occurred in the concrete, all of the mortar cylinders split at the yield point.

2. Examination of the enamel surfaces shows no morphological change in the cement grains that indicates firing the cement-enamel mixture on the surface of steel altered the cement.

3. The microanalytical investigation indicates that the changes in chemical composition of the coatings were consistent with normal hydration reactions in portland cement.

4. X-ray diffraction investigations of the unhydrated enamel showed the firing of the mixture to fuse the glass to the steel surface did not convert all of the gypsum to anhydrite. All of the phases needed for normal concrete setting reactions to occur were present in the cement-enamel coating.

5. The crystalline phases in the cemented glass-cement coating sample were consistent with the observation that mortar was still bonded to the enamel after the metal and mortar separated.

6. Additional work with other types of silicate cements should demonstrate similar or improved bond strengths can be obtained using glass-cement bonding coatings.

5. REFERENCES

1. Bentur, A. and Mindess, S., Fibre Reinforced Cementitious Composites, *Elsevier Applied Science Publishers,* London, UK, 1998.

2. Swamy, R. and Barr, B., Fibre Reinforced Cement and Concretes: Recent Developments, *Spon Press,* UK, London, 1990.

3. Al Khalaf, M. and Page, C., Steel mortar interfaces: Microstructural Features and Mode of Failure. *Cement and Concrete Res.* 9, 1979, pp. 197-208.

4.   Mindess, S., Mechanical Properties of the Interfacial Transition Zone: A Review. In Buyukozturk, O. and Wecharatana, M. Interfacial Fracture and Bond. SP-156, Interface Fracture and Bond. *American Concrete Institute*, Detroit, 1991, p. 1-9.

5.   Fu, X., and Chung, D., Effects Of Water-Cement Ratio, Curing Age, Silica Fume, Polymer Admixtures, Steel Surface Treatments And Corrosion on Bond between Concrete And Steel Reinforcing Bars. *ACI Materials. Journal*, 95 (6), 1998, pp.725-734.

6   Li, Z. Xu, M., Chung, C., Enhancement of Rebar (Smooth Surface)--Concrete Bond Properties By Matrix Modification And Surface Coatings. *Magazine of Concrete Research*, 50(1), 1998, pp. 49-57.

7.   Thangavel K., Rengaswamy, N, Balakrishnan, K. Influence Of Protective Coatings On Steel--Concrete Bond. *Indian Concrete Journal*, 69(5), 1995, pp. 289-293.

8.   Lea, F., The Chemistry of Cement and Concrete Third Edition, *Chemical Publishing Co., Inc.*, New York, NY, 1971.

9.   Neville, A., Properties of Concrete, Third Edition, *Pitman Publishing Ltd.*, London, 1981.

10.   Taylor, H. Cement Chemistry, 2nd edition, *Thomas Telford Publishing*, London, 1997.

11.   Fu, X. and Chung, D., Sensitivity of The Bond Strength to the Structure of the Interface between Reinforcement and Cement, and the Variability of this Structure. *Cement and Concrete Research*, 1998, 28(6), pp. 787-793.

12.   Bentur, A. and Cohen, M., Effect of Condensed Silica Fume on the Microstructure of the Interfacial Zone in Portland Cement Mortars. *Journal of the American Ceramic Society*, 70(10), 1987, pp. 738-743.

13.   Fu, X., and Chung. D., Improving The Bond Strength of Concrete to Reinforcement by Adding Methylcellulose to Concrete. *ACI Materials Journal*, 95(5), 1998, pp.601-608.

14.   Mazkewitsch, A. and Jaworski, A., The Adhesion Between Concrete and Formwork. pp. 67-72 in Salle, H., Adhesion between Polymers and Concrete, *Chapman and Hall*, New York, NY, 1986.

15.   Mlodecki, J., Adhesion of Polymer Modified Concrete and Plain Concrete to Steel In Moulds and in Reinforced Concretes. pp. 55-64 in Salle, H., Adhesion between Polymers and Concrete, *Chapman and Hall*, New York, NY, 1986.

16.   Society of Manufacturing Engineers, Porcelain Enameling, *Society of Manufacturing Engineers*, Dearborn, MI, (undated).

17.   Danielson, R. and Wolfram, H. Vitreous Enamels for Metals in Washburn, E. W. International Critical Tables of Numerical Data for Physics, Chemistry and Technology. Vol. 2, pp. 114-117, *Knovel Corp.*, Norwich, NY, 2003.

18.   Day, D., Carrasquillo, M., Weiss, C., Sykes, M., Baugher, E., and Malone, P. Innovative Approaches To Improving The Bond Between Concrete And Steel Surfaces. Proceeding of the 25th Army Science Conference, Orlando FL. 2006.

19.   American Society for Testing and Materials, Standard Method for Determining Compressive Strength of Hydraulic Mortars, ASTM Designation C 109, *American Society for Testing and Materials*, West Conshohocken, PA, 1999.

20.   Kosmatka, S. and Panarese, W., Design and Control of Concrete Mixtures, *Portland Cement Association*, Skokie, IL, 1990.

21.   American Society for Testing and Materials, Standard Test Method for Comparing Bond Strength of Steel Reinforcing Bars to Concrete Using Beam-End Specimens. ASTM Designation A 944-99, *American Society for Testing and Materials*, West Conshohocken, PA, 1999.

22. Maage, M., Fibre Bond and Friction in Cement and Concrete. RILEM Symp. On Testing and Test Methods of Fibre Cement Composites, *The Construction Press*, Hornby England, 1978, Paper 6.1, pp. 329-336.
23. Hackler, C., Koenigstein, M., and. Malone, P., The Use of Porcelain Enamel Coatings on Reinforcing Steel to Enhance the Bond to Concrete, Materials Science and Technology 2006 Conference and Exhibition, Conference Proceedings. *American Ceramic Society*, Westerville, OH.

IMPORTANCE OF RHEOLOGY IN ENAMEL APPLICATION

Michael Sierocki
AO Smith Protective Coatings Division
Florence, KY

This paper will first discuss the benefits and deficiencies of various rheology measurement techniques. Discussion leads to how certain measurable components of rheology are directly related to various application methods of a porcelain enamel, then how we as enamel scientists can benefit the applicator by reducing defects and costs.

The discussion of rheology in this paper is regarding the properties of the porcelain enamel slip rather than the properties of the melt.

DEFINING RHEOLOGY

What is rheology? The classical definition of rheology is the study of flow and deformation of matter. To the formulator of porcelain enamels this typically means studying slip flow to optimize physical and chemical properties per customer requirements. To the applicator, rheology typically has an additional meaning; that is to optimize processing and product consistency while minimizing costs. The rheology characteristics and consistency in a porcelain enamel slip can mean the difference in making or losing money.

In rheology there are some differences in terminology depending on what business you are in, for instance the word "slip." In certain coating applications slip is the term referring to friction of a surface. In the porcelain enameller's case it also means the wet slurry applied onto a substrate before firing. The following definitions will be used for the purposes of this paper.

Newtonian: a characteristic of a material that exhibits no change in viscosity under changes in shear stress or shear rate
Slip: the fluid mixture of water and ceramic components prior to application on a substrate
Shear Stress: the stress from external components exerted on the slip
Shear Rate: the rate of movement of the slip due to external stresses; shear rate can be directly related to any application or manufacturing process (i.e. pumping, spraying, settling, mixing, sagging, leveling, milling, etc.)
Viscosity: The resistance to flow
Shear Thinning: reduction of viscosity with increasing stress
Thixotropic: a shear-thinning fluid that recovers viscosity during rest
Flow: viscosity profile of the slip under stress
Yield Stress: Minimum amount of stress required to cause permanent deformation of the slip
Recovery: Viscosity profile of the slip at rest, typically directly after stressing the slip; can be loosely translated to set

CONSIDERATIONS

There are very important areas of variability to consider when measuring any aspect of rheology. Among them are temperature, shear history, sampling and test repeatability, usually because of human variance. Controlling these variables will increase the repeatability and reproducibility of your measurements.

Temperature fluctuations in the application area can have unfavorable effects on the slip. For example if the temperature is higher (say summertime) the slip may have lower viscosity at a particular part of the process. It may also recover faster, changing the results for pickup or slump, misleading the operator to believe the material is alright to use.

Sampling can definitely cause variance in rheology measurements. Perhaps one sample is taken directly after a mixing or milling the slip and another may be taken a few minutes after. If both samples are from the top, which is very typical in large vessels, any settling or phase separation that has occurred before sampling will alter the measurements taken with the samples.

Shear history is also a key component in accurate rheology measurements. Time is an important variable for all thixotropic fluids. Remember that a thixotropic material is one that recovers its viscosity after shearing. Therefore the history of shear before analysis will determine the state of the sample; whether the sample has somewhat recovered, mostly recovered, fully recovered, etc.

Human variance is a variable to consider in any measurement; rheology measurements are no exception. The biggest problem with human variance is that while it is a variable on its own in the actual running of the test, it also contributes to other variables mentioned, like sampling and shear history. Depending on the analyst, variability of the end result can be substantial.

To reduce variability, increase control of your procedures and processes. Performing an analysis of variance (ANOVA) and gage R&R study will outline the variability of your operators as well as the gages used to measure the rheology.

VARIOUS MEASUREMENTS OF RHEOLOGY

Slump

A slump gage consists of a cylinder and a plate. The slump of a material is determined by filling the known cylinder with your porcelain enamel slip then lifting the cylinder, allowing the slip to flow over the plate. The measurement taken is the diameter of the resulting pool.

Slump measurements are a widely accepted measurement of rheology for porcelain enamel slips. ASTM C143/C143M-05A is the test typically followed, however it was designed for concrete. With that being said, consider the differences in application of concrete and porcelain enamels. Cement is typically poured into a void or troweled into place, then allowed to sit and cure. Porcelain enamel application typically occurs by dipping and draining the part into a bath of the slip, allowing the slip to be poured over the part, or spraying the slip onto the part.

The main problem with slump measurement is that the methods of porcelain enamel application usually do not even resemble that of concrete. Important factors of application such as film thickness, film uniformity, spray or dip viscosity, settling, etc. are not easily related. The best relationship slump has with the application process is gravity related, like discharging a wet mill of the enamel and draining a part after the enamel was poured on or dipped.

Another problem with slump is operator variability. Unless the slump gage is automated there are variances like the rate and angle of the cylinder as it is lifted and time variations regarding how long the material sits in the cylinder before it is lifted.

Pickup

Pickup is performed a couple different ways. Some applicators prefer using a cylinder; others prefer a plate of sheet metal. Either way the same basic procedure is used: tare the uncoated plate/ cylinder on a scale, dip the plate/ cylinder in a bath of the slip, pull out of the bath and let drain, re-weigh. The mass of slip picked up divided by the total surface area provides a good estimate to coating usage.

Further calculation using specific gravity can give you average film thickness as well; however this calculation is averaged over the whole part and therefore can not provide information about the uniformity of the coating. Sagging, leveling and other typical film-build problems encountered at application are not being considered in this test.

Viscometers

Viscometers, such as a Brookfield or Stormer type, have the ability to measure viscosity over a range of shear. They do this by measuring the torque output of the motor to move the measuring geometry at a certain rotational speed. This does help in characterizing the porcelain enamel slip at various points of application or manufacturing; however these types or viscometers are still quite limited.

While you may be able to determine the viscosity within a narrow band of shear, there is still some valuable information that can not be measured, such as viscosity at settling, recovery, yield, and viscosity at high-shear application. The viscometer is unable to measure the torque at a rotational rate slow enough to mimic settling. They also can not rotate fast enough to mimic spray application. Recovery and yield stress are at best poorly estimated since there needs to be rotation for a viscometer to measure viscosity – once rotation begins, yield has already occurred and recovery can not occur!

Rheometers

The rheometers discussed in this section are of the research grade models available from manufacturers like TA Instruments, Paar Physica and Bohlin (Malvern). The rheometer used for the purpose of this paper was the TA Instruments AR-G2.

These rheometers are rotational like the viscometer; however they use intricate bearings that allow for virtually "frictionless" movement of the measuring geometry. Therefore they are capable of accurately measuring viscosity at extremely low stresses, causing minimal movement of the fluid. The result is a much more accurate analysis of characteristics like recovery, yield or viscosity at settling.

Since we can now measure these sensitive characteristics, the slip can be formulated to meet more detailed application or manufacturing requirements. The applicator can also use this information to better streamline specific parts of their process, reducing costs and improving product quality.

There are other important properties that these rheometers can also measure, like modulus and other viscoelastic properties, however they are beyond the scope of this paper and will not be discussed.

RHEOLOGICAL MEASUREMENT DATA

The following data was generated by the author using a TA Instruments AR-G2. The measurement geometry is standard DIN concentric cylinders.

Sample information
  Slip 1 – Initial formula, 6 ½ inch slump
  Slip 2 – Modified formula, 4 ½ inch slump
  Slip 3 – Final formula, 6 inch slump

Measuring Yield

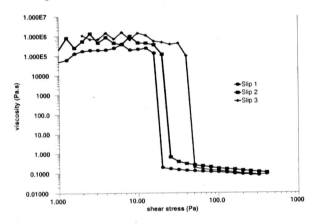

In this first analysis a stress is applied to the measuring geometry and the viscosity is calculated at steady state. The stress then ramps up and another measurement is taken, and so on.

As you can see all three samples have similar properties, in that they exhibit a high level of viscosity until a certain amount of stress is applied. After this yield stress the viscosity drops to another level. It is quite obvious to the analyst that Slip 3 has the highest yield stress, with some improvement of Slip 2 over Slip 1.

Recall that yield stress is the minimum amount of stress needed to cause the material to flow. Regarding these samples Slip 3 will require the most stress to start flowing. In the example of a sprayed or dipped part, Slip 3 will be the most resistant to dripping once the slip has set, allowing for rougher handling of the part prior to firing. Slip 1 will be more susceptible to flowing after the part has been coated, allowing more dripping to occur.

Measuring Flow

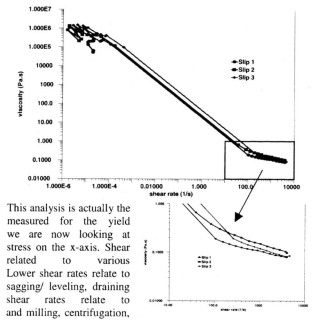

This analysis is actually the measured for the yield we are now looking at stress on the x-axis. Shear related to various Lower shear rates relate to sagging/ leveling, draining shear rates relate to and milling, centrifugation, same data that was stress analysis, however shear rate rather than rates can be directly application processes. processes like settling, and pipe flow. Higher processes like dispersing pumping and spraying.

As you can see the samples appear to have very similar flow properties. Upon closer examination, however, Slip 1 and Slip 3 have nearly the same viscosity at high shear rates. This means they will spray at nearly the same volumetric flow rate, have similar mixing viscosity and can be pumped in the same manner. Slip 2 will have higher viscosity at these processes, meaning harder to mix and more pressure will be required to spray at the same flow rate.

Measuring Recovery

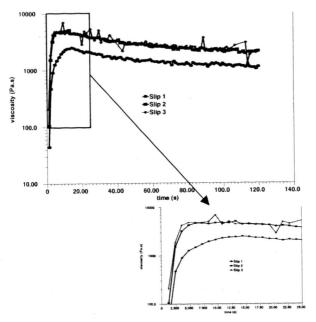

For this analysis we are looking at how the viscosity of the samples recovers over time. To perform this, the rheometer is programr. ed to apply a small enough stress that it will minimally disturb the sample; however it still can measure viscosity.

Notice how the recovery is much improved for Slip 2 and Slip 3 over Slip 1, meaning that the coating will have better dripping and sagging resistance.

RESULTS OF RHEOLOGICAL ANALYSIS

So, which slip sample turned out to be the best for the applicator? Let us review.

Slip 1: Initial formula, drips form at bottom flange and disrupt the fitting.
Slip 2: Rheology modification of Slip 1, dripping is much less of a factor (less re-worked parts), however can not build wet film as effectively as Slip 1.
Slip 3: Further modification of Slip 2. Dripping is still much improved over Slip 1, however it sprays much better and wet film thickness is again achievable.

Obviously the applicator was happiest with Slip 3, which causes less re-work than the original formulation for the applicator.

BENEFITS OF OPTIMIZED RHEOLOGY

In the example analysis it was shown how certain aspects of application are improved when rheology is optimized. Although it was shown how the rheology improves each sample, the applicator was more interested in the bottom-line result: less rework! However it should be known that reduction of scrap or re-work is just the immediate effect of optimized rheology. Beyond these savings are reduced service issues and reduced adjustments to an unstable slip, just to name a few.

The reduced service issues are due to more uniform and consistent application of the coating. Defects such as burn-off and copperheads can be due to thin enamel, drain lines and orange peel can be caused by poor recovery, sagging can be caused by either poor recovery or thick coating, and spalling can be caused by excessive thickness.

THE BOTTOM LINE

This presentation has shown various methods of rheology measurement and characterization, emphasized the importance of properly measuring to benefit application and went through a real-life example of how rheology helped an applicator.

And the bottom line is: Optimized Rheology = Reduced Costs

Author Index